Thermal Behavior of Photovoltaic Devices

Olivier Dupré · Rodolphe Vaillon
Martin A. Green

Thermal Behavior of Photovoltaic Devices

Physics and Engineering

Springer

Olivier Dupré
Centre for Energy and Thermal Sciences of Lyon, CNRS, Univ Lyon, INSA-Lyon
Université Claude Bernard Lyon 1
Villeurbanne
France

Rodolphe Vaillon
Centre for Energy and Thermal Sciences of Lyon, CNRS, Univ Lyon, INSA-Lyon
Université Claude Bernard Lyon 1
Villeurbanne
France

and

Radiative Energy Transfer Laboratory, Department of Mechanical Engineering
University of Utah
Salt Lake City
USA

Martin A. Green
Australian Centre for Advanced Photovoltaics, School of Photovoltaic and Renewable Energy Engineering
University of New South Wales
Sydney
Australia

ISBN 978-3-319-49456-2 ISBN 978-3-319-49457-9 (eBook)
DOI 10.1007/978-3-319-49457-9

Library of Congress Control Number: 2016958462

Printed on acid-free paper

This Springer imprint is published by Springer Nature
The registered company is Springer International Publishing AG
The registered company address is: Gewerbestrasse 11, 6330 Cham, Switzerland

Preface

Although long known that temperature negatively affects the performances of PV devices, research and engineering in photovoltaics has been mainly focused on reducing the optical and electrical losses. This book is the result of several years of research dedicated to understand better the thermal behavior of photovoltaic devices. The work on this topic was triggered by simple questions such as: Why does temperature negatively affect the efficiency of most photovoltaic systems? Is it possible to engineer the temperature sensitivities of solar cells? If it is not possible to make their performance completely independent of temperature, what can be done to minimize their operating temperatures?

In this book, we discuss the different temperature-driven phenomena that are important in the field of photovoltaics. We provide detailed physics-based explanations of the mechanisms involved and review different solutions and strategies to mitigate the temperature-induced losses. The original perspective we have taken in this book to describe the fundamental principles of photovoltaic conversion has its origins in the diversity of our backgrounds. Throughout the book, we introduce a novel approach to optimize the design of photovoltaic devices where thermal criteria are integrated. Because the thermal behavior of a photovoltaic system is deeply intertwined with its electrical and optical/radiative characteristics, the optimum set of design parameters is always a function of the operating conditions, which depend on the location and the mounting configuration of the system. This means that photovoltaic devices can be designed to maximize their energy yield in specific conditions (climates, installations, etc.). The present book provides models and concepts to guide researchers and engineers towards the development of this kind of optimization that goes beyond the standard test conditions.

We are very grateful to all the persons that have motivated, through direct contacts or publications, our research and those who helped during the preparation of the book. Yannick Riesen and Ricky Dunbar deserve special mention for stimulating discussions about the specific thermal behavior of silicon heterojunction and perovskite-based solar cells. We also thank particularly Anne Gerd Imenes, Sarah Kurtz and David Moser for kindly providing data of photovoltaic modules installed all around the world.

Rodolphe Vaillon was hosted by the Department of Mechanical Engineering at the University of Utah when the present book was prepared. The financial support from the College of Engineering (W.W. Clyde Visiting Chair award) is acknowledged.

Villeurbanne, France — Olivier Dupré
Salt Lake City, USA — Rodolphe Vaillon
Sydney, Australia — Martin A. Green

Contents

Nomenclature

Fundamental Constants

c	Speed of light in vacuum ($\approx$2.99792458 10^{8} m s^{-1})
h	Planck constant ($\approx$6.62607004 10^{-34} J s)
k	Boltzmann constant ($\approx$1.38064852 10^{-23} J K^{-1})
q	Elementary charge ($\approx$1.60217662 10^{-19} C)
σ	Stefan–Boltzmann constant σ ($\approx$5.67037321 10^{-8} W m^{-2} K^{-4})

Latin Symbols

D	Diffusion coefficient (m^2 s^{-1})
E	Photon energy (eV)
$E_{c/v}$	Conduction/valence band edge (eV)
$E_{Fn/p}$	Quasi-Fermi level of the electrons/holes (eV)
E_g	Energy bandgap (eV)
EQE	External quantum efficiency (–)
ERE	External radiative efficiency (–)
f	Fraction of photons (–)
G	Parameter of interest
h	Heat transfer coefficient (W m^{-2} K^{-1})
I	Current (A)
J	Current density (A m^{-2})
L	Diffusion length (cm)
n	Photon flux density (m^{-2}) or diode ideality factor (–) or electron concentration (cm^{-3})
N	Doping concentration or density of states (cm^{-3})
p	Hole concentration (cm^{-3})
P	Power loss per unit area not being a heat source (W m^{-2})

PFD	Photon flux density (m^{-2})
$PFD_{bb(T)}$	Photon flux density emitted by a blackbody at temperature T (m^{-2})
Q	Component of the heat generated in the device per unit area, heat flux (W m^{-2})
R(*E*)	Spectral reflectance of the PV device (–)
R	Resistance (Ω)
S	Entropy (J/K)
T(*E*)	Spectral transmittance of the PV device (–)
T	Temperature (K or °C)
V	Voltage (V)
V_{max}	Maximum voltage (V)
W	Cell thickness (m) or useful work
X	Concentration factor (–)

Greek Symbols

β	Temperature coefficient (ppm K^{-1} or % K^{-1})
η	Conversion efficiency (–)
Δ	For specifying a current (ΔJ) or a voltage (ΔV) loss with respect to the maximum
$\varepsilon(E)$	Spectral emittance of the PV device (–)
Ω	(projected) solid angle
Π	Peltier coefficient (V)
μ	Chemical potential (eV)
τ	Carrier lifetime (s)
ξ	Parameter closely related to the product of carriers np introduced in (Green 2003)
γ	Parameter that includes the temperature dependences of several parameters determining J_0

Subscripts

0	In the dark (applies to the current density)
1sun	Under one sun illumination
abs	Applies to the (projected) solid angle within which the cell absorbs radiation from the source
am	Refers to the angle mismatch loss
amb	Ambient

bb	Refers to the case where the illumination (or the cell) radiation is modelled by a blackbody (Planck function)
below E_g	Refers to the sub-bandgap loss
c	Refers to the cell
Carnot	Carnot fundamental loss
cond, conv, rad	Apply to the heat fluxes exchanged by the cell with its surrounding by conduction, convection, radiation
current	For the fraction of photons that participate to the current
ΔJ, ΔV	Apply to the heat source components associated to a current loss or a voltage loss
e	Environment with which the device exchanges thermal radiation
emit	Applies to the (projected) solid angle within which the cell emits radiation
ex	For the fraction of emitted photons that exit the cell
G	Applies to the parameter G of interest
i	Intrinsic (applies to the carrier concentration without doping)
ib	For the fraction of photons absorbed through the interband process
inc	Incident upon the cell
MPP	Refers to the maximum power point
oc	Refers to the open circuit configuration
par	For the fraction of photons absorbed through parasitic processes
rad	In the radiative limit
s	Refers to the Sun (equivalent blackbody temperature only) or series (for resistances)
sc	Refers to the short circuit configuration
sh	Shunt (for resistances)

Abbreviations

AM	Air mass
ARC	Anti reflection coating
BC	Boundary condition
BSF	Back surface field
c-Si	Crystalline Silicon
CdS	Cadmium Sulfide
CdTe	Cadmium Telluride
CIGS	Copper Indium Gallium Selenide
CPV	Concentrated photovoltaics
EG-Si	Electronic grade silicon
EQE	External quantum efficiency
ERE	External radiative efficiency
EVA	Ethylene-vinyl acetate
FCA	Free carrier absorption
FF	Fill factor
FZ	Float zone
GaAs	Gallium Arsenide
Ge	Germanium
HIT	Heterojunction intrinsic thin layer
InP	Indium Phosphide
IR	Infrared
LED	Light-emitting diode
MPP	Maximum power point
NF-TPV	Near-field thermophotovoltaic
NOCT	Nominal operating cell temperature
NRR	Nonradiative recombination
PERC	Passivated emitter and rear cell
PERL	Passivated emitter with rear locally diffused
PFD	Photon flux density
PR	Photon recycling

PV	Photovoltaic
RR	Radiative recombination
SHJ	Silicon heterojunction
Si	Silicon
SiN	Silicon Nitride
SQ	Shockley–Queisser
SRH	Shockley–Read–Hall
STC	Standard test conditions
STPV	Solar thermophotovoltaic
TC	Temperature coefficient
TPV	Thermophotovoltaic
UV	Ultraviolet
ZnS	Zinc Sulfide

Chapter 1
Thermal Issues in Photovoltaics and Existing Solutions

Abstract This chapter introduces the different effects of temperature on the performances of photovoltaic (PV) devices. Efficiency of the vast majority of photovoltaic devices drops when their temperature rises. An augmentation of temperature has other effects on certain types of PV devices such a promoting the regeneration of amorphous silicon cells. High device temperature is also a factor favoring different degradation mechanisms such as potential induced degradation (PID). All of these temperature-induced effects have important implications for the photovoltaic industry. Indeed, two different cells or modules with the same rated power in the standard test conditions (STC), i.e. 1000 W m^{-2} of AM1.5 illumination and a cell temperature of 25 °C, may produce different electrical powers under real outdoor conditions and thus have different energy yields. To forecast accurately the energy production of PV plants, it is thus necessary to predict their operating temperatures. This chapter presents an overview of the models developed towards that end. Then, different strategies to reduce the operating temperature of PV devices are presented. Several active and passive cooling methods are introduced. The specificities of hybrid photovoltaic/thermal (PV-T) systems, which generate both heat and electricity, and building integrated photovoltaics (BIPV), where the PV modules are part of the building envelope, are discussed. Finally, the potential of radiative cooling and other thermal design approaches of solar cells and modules are evaluated.

1.1 The Effects of Temperature on the Performances of Photovoltaic Devices

The performance of a photovoltaic device is naturally dependent on the temperature at which it operates since most of the physical mechanisms involved in the photovoltaic conversion process are temperature-dependent. Without describing the underlying physics, which is done in Chap. 2, this chapter introduces the different temperature-induced effects relevant for photovoltaic applications.

O. Dupré et al., *Thermal Behavior of Photovoltaic Devices*,
DOI 10.1007/978-3-319-49457-9_1

1.1.1 *Reversible Decrease of PV Performances with Temperature*

It is widely known that the conversion efficiency of solar cells decreases when their temperature rises. For most type of solar cells,[1] this decrease is linear with temperature (see Fig. 1.1). Consequently, this phenomenon can be accurately characterized by a single value commonly referred to as temperature coefficient that corresponds to the rate of variation with temperature.[2] The negative temperature coefficient of PV conversion efficiency is mainly due to a strong negative temperature dependence of the open-circuit voltage of PV devices (Prince 1955; Green 2003). The physical origin of this phenomenon is that the rates of most recombination mechanisms increase with temperature. However, recombinations alone are not sufficient to explain the complex temperature sensitivities of solar cells. In certain cases for example, the temperature dependence of the efficiency of a solar cell is mainly driven by that of a limiting transport mechanism. Chapter 2 provides a detailed explanation of the physics driving the temperature coefficients of actual solar cells.

Early investigations on the temperature dependence of solar cell conversion efficiency highlighted the importance of the bandgap[3] of the semiconductor used in the device (Halsted 1957; Wysocki and Rappaport 1960). Indeed, solar cells made from different semiconductors[4] exhibit different temperature sensitivities (Fig. 1.1). Also, because the semiconductor bandgap is not the only parameter involved, solar cells made from the same semiconductor may exhibit different temperature dependences. This is illustrated in Fig. 1.1 for silicon cells and detailed explanations are provided in Sect. 2.3.

Figure 1.1 illustrates why the dependence of conversion efficiency on device temperature is especially important for the photovoltaic industry. Two different cells or modules with the same rated power in the Standard Test Conditions (STC), i.e. 1000 W m^{-2} of AM1.5 illumination on a cell at 25 °C, may produce different electrical powers under real outdoor conditions. The example of the range of operating temperatures in Norway is exceptional because the average operating temperature in terms of energy production is around 25 °C (see Fig. 1.2). This low average operating temperature is due to the fact that the climate at the location of the module (Norway) is especially cold. In most locations, where the solar irradiance is important over the year, the average operating temperature is significantly above 25 °C (see for example Fig. 1.3) and the STC are not representative of the average operating

[1]With the exception of certain solar cells such as those made of amorphous silicon. Detailed explanations are given in Sect. 1.1.2.

[2]See the definition in Sect. 2.1.

[3]Bandgap or energy bandgap corresponds to the energy separating the valence band from the conduction band in a semiconductor.

[4]a-Si: amorphous silicon. CdTe: cadmium telluride. c-Si: crystalline silicon. CIGS: copper indium gallium selenide. SHJ: silicon hetero-junction.

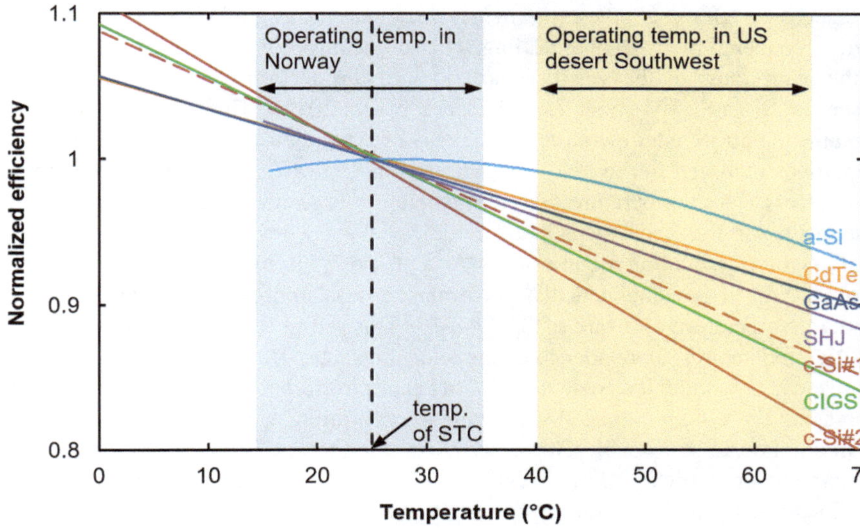

Fig. 1.1 Conversion efficiency normalized to its value at 25 °C as a function of temperature for different PV cell technologies. Range of operating temperatures for ~80% of the annual energy production in Norway and in the US desert Southwest. Data for the efficiencies from Virtuani et al. (2010), Dupré et al. (2015). Data for the operating temperature from First Solar (2014) and from the Energy Materials Group at the University of Agder, Norway (see Figs. 1.2 and 1.3)

conditions (Kurtz et al. 2011; Strevel et al. 2012; Moser et al. 2013). Thus, the knowledge of the temperature coefficients (and the dependences upon irradiance) is crucial to estimate the energy yield of PV plants. Consequently, in the past decades, research related to the thermal behavior of solar cells aimed at:

- predicting the operating temperatures of PV plants in order to assess potential energy yield or forecast the power production of existing PV installations. An overview of the models available to date is presented in Sect. 1.2.
- mitigating the temperature-induced losses by reducing the operating temperature. Different strategies suggested towards that end are introduced in Sect. 1.3.

1.1.2 Thermal Annealing and Staebler-Wronsky Effect in Amorphous Silicon Solar Cells

Amorphous silicon (a-Si) based solar cells have a peculiar thermal behavior resulting from specific mechanisms such as light soaking and thermal annealing (Carlson et al. 2000). The metastable light-induced degradation of hydrogenated amorphous silicon (a-Si:H) is referred to as the Staebler-Wronsky effect after its discoverers (Staebler and Wronski 1977). While the physics involved remains a topic of discussion (Branz 2003), experiments show that the degradation can be, at least partly, recovered by thermal annealing. This annealing-promoted regeneration effect is the main cause of the seasonal variations of the performance ratio[5] commonly observed for a-Si PV modules (Makrides et al. 2012). This phenomenon also explains the positive temperature coefficients of amorphous silicon modules that are sometimes recorded in the field (Fischer et al. 2009; Moser et al. 2013; Silverman et al. 2014). This temperature-driven regeneration is so important that several publications report that backside isolated modules have a better energy yield over a year than classic freestanding modules (Kondo et al. 1997; Fischer et al. 2009). In Fischer et al. (2009), it is calculated that performance ratio improves together with peak module temperature at a rate of 0.87% K^{-1}. This is linked to the very strong and positive temperature dependence of the annealing rate. These somewhat surprising results lead to wonder if a-Si devices could benefit from specific thermal designs (e.g. increasing the usually parasitic sub-bandgap absorption, see Chap. 3).

In addition to the Staebler-Wronsky effect, the thermal behavior of a-Si devices exhibits other particularities. These specific temperature dependences and their implications for device optimization are discussed in Sect. 4.4.

1.1.3 Temperature and Module Degradation

A “rule of thumb” in studying any chemical reaction is that, near room temperature, reaction rates frequently double for each 10 °C increase in temperature (Pauling 1988). Assuming Arrhenius behavior, this corresponds to an activation energy in the 0.5–0.6 eV range. Applying this “rule of thumb” to the degradation process determining the life of a solar module, module life could be considerably extended with even a modest reduction in its operating temperature. A 1 °C decrease in this temperature could extend module life by about 7%, or 2 years. This is an additional advantage, not widely accounted for, of the different strategies that reduce the operating temperature of PV devices (see Sects. 1.3 and 3.5).

This “rule of thumb” prediction can be checked against the results from more detailed studies. One specific module degradation mode attracting much attention recently is potential induced degradation (PID) (Hoffmann and Koehl 2012). To reduce wiring losses and costs, modules in large photovoltaic systems generally are operated at large DC voltages, with 1000 VDC presently common with a trend now towards 1500 VDC operation. While many cells in the system will be at potentials approaching these values, metallic parts of the system such as module frames and support structures need to be kept at ground potential, for safety reasons.

[5]The performance ratio (PR) of a PV module is defined as the ratio between its actual energy yield and its maximum theoretical energy yield given by the incoming irradiance and the module nominal efficiency.

Consequently, large potential differences can exist between parts of the system that are in close physical proximity.

For a pristine system, this does not pose a problem since the cells are packaged in electrically well-insulated modules including, in general, an insulating glass coversheet. However in the field, the front of the glass sheet can become partially conducting, particularly when deposits have built up on the glass surface and the surface is moist. This can result in the transfer of the near-ground potential to the glass surface immediately above a cell that is operating at a very high positive or negative potential. The high corresponding electric field strength results in leakage currents between the cell and ground, with these supported by ionic motion in the normally insulating layers separating the cells from the top glass surface. These ions can damage the module such as well documented in the case of sodium ions, with these able to enter the silicon cell surface and shunt the cell (Naumann et al. 2014).

Potential induced degradation rates are accelerated by both high temperature and humidity. A controlled study involving both laboratory and field data provides a characterization of their combined effect. At fixed humidity, increasing temperature increases the leakage currents responsible for the module degradation with an activation energy of 0.75 eV (Hoffmann and Koehl 2012). This is slightly higher than the typical range previously mentioned, corresponding to a doubling of the leakage for every 7 °C temperature rise, although the temperature dependence of the process is still not far from that described by the standard "rule of thumb".

In terms of other observed degradation modes, the most challenging acceptance test for solar modules is generally considered to be the 85% humidity, 85 °C damp heat test, where less than 5% module degradation after 1000 h testing is required, for the module to pass. Under this test, one degradation mechanism for silicon modules is moisture ingress, resulting in corrosion of the metal contacts and/or their interface with the silicon, beginning at extremities of the metallization (Peike et al. 2013). From extensive measurements of degradation rates under different temperature and humidity combinations, an activation energy of 0.49 eV is determined in Kim et al. (2013). This is slightly lower than the "rule of thumb" value, with degradation rates for a given humidity doubling for every 12 °C temperature rise near room temperature.

The two specific studies outlined above suggest that the "rule of thumb" may be a reasonable starting point for projecting module degradation rates, regardless of the details of the degradation process. Variations might be expected when temperature changes introduce phase changes in a module component if, for example, this component melts.

1.2 Predicting the Operating Temperature and Energy Yield of PV Installations

The power output of a PV device is function of the radiation incident upon it and also on the device operating temperature via the temperature coefficient. Thus, to assess the potential energy yield and in turn the economic viability of a PV installation, it is necessary to be able to predict its operating temperature. Short term forecasting of power outputs is also required for electricity suppliers to continuously balance the grid and optimize the distribution of the various electricity sources.

Figure 1.2 shows the temperature distribution of module temperature under real operating conditions for a monocrystalline silicon module installed in Southern Norway. Figure 1.2a illustrates that the module temperature is generally higher than the ambient. This is due to the fact that a large fraction of the incident solar energy is converted into heat within the device (see Chap. 3), and depending on the thermal conditions (ambient temperature, wind speed and orientation, etc.) and the thermal design of the installation (ability to release heat by conduction, convection and radiation), the module temperature rises above the ambient. The average temperature over energy production (shown in Fig. 1.2b) is always larger than the average temperature over time (shown in Fig. 1.2a). This happens because energy production is more or less proportional to irradiance which contributes to heating the module. In other words, a large fraction of energy production takes place at periods of high irradiation which coincides with periods when the operating temperature is the largest. Thus if a larger level of incoming irradiation has beneficial impacts on energy production, the concomitant larger device temperature somewhat reduces this beneficial effect (see also Fig. 1.5). Anyhow, to calculate PV energy yield, average temperature over energy production should be considered as the most representative averaged temperature.

Interestingly, in this relatively rare scenario of cold climate (Norway), the distribution in terms of energy production is symmetric about an average close to the temperature in the standard test conditions (STC) which is 25 °C. Hence, in this case, optimization of the performances of modules in STC seems appropriate.

Figure 1.3 shows the temperature distribution of the back surface of CdTe modules located in a power plant in the US desert southwest. It illustrates that in operation, the temperature of PV modules in a warm climate is much larger than the 25 °C of the STC.

In the same way, Fig. 1.4 shows the modeled temperature distribution, in hours per year, of modules at different locations in the world (Kurtz et al. 2011). These examples illustrate once again the impact of local climate on the distribution of module operating temperature (see also Koehl et al. 2011). For a given climate, the ability of a module to dissipate the internally generated heat and in turn reduce its operating temperature also depends on the mounting configuration (open-rack, roof mounted, insulated back, tracker, etc.). This is illustrated by the difference in temperature distributions of modules mounted on open racks (Fig. 1.4a) and of modules with insulated backs (Fig. 1.4b). It is not surprising that modules with an insulated back reach higher temperatures than those mounted on open racks. Note that the temperature distributions depicted in Fig. 1.4 include nighttime (unlike the data presented in Figs. 1.2 and 1.3). Also, as previously discussed, the temperature distribution in terms of energy production would be quite different.

The ability of different modules to dissipate heat can be quantified by the so-called nominal operating cell temperature (NOCT). It is the temperature reached by the open-circuited cells in a module for a specific mounting configuration (open-rack) in given thermal and irradiance conditions (wind velocity of 1 m s^{-1}, an ambient air temperature of 20 °C and under a 800 W m^{-2} illumination) (Ross

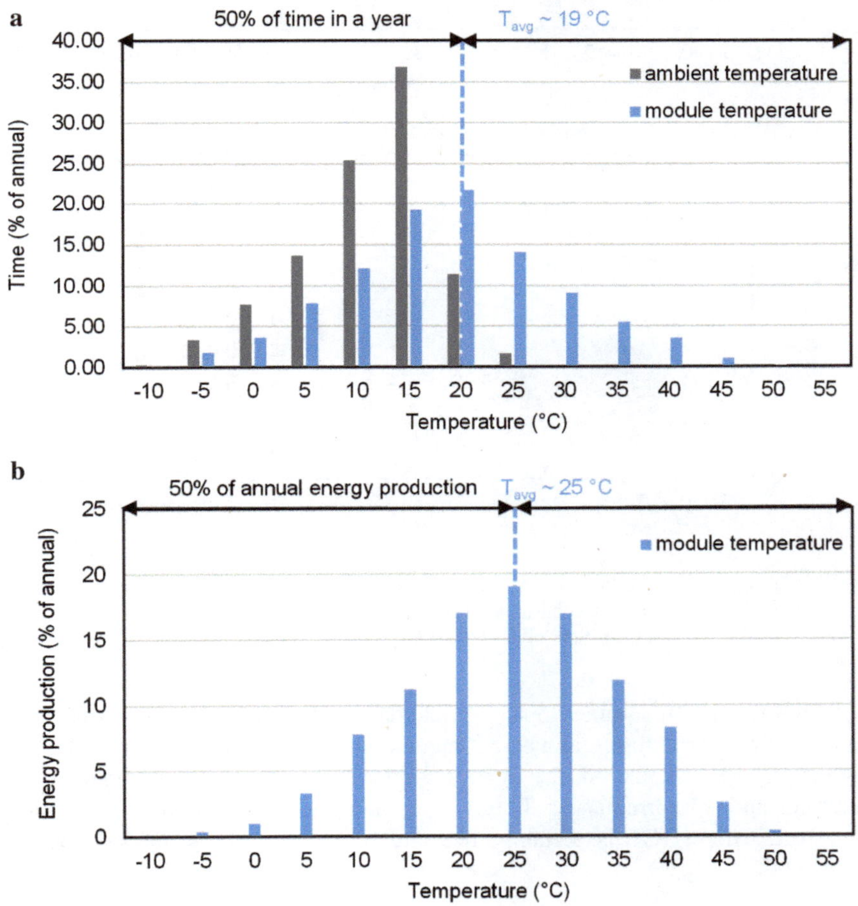

Fig. 1.2 Temperature distribution under real operating conditions of a mono-crystalline silicon module installed in Southern Norway over: **a** time in a year; **b** annual energy production. Figure created with data provided by the Energy Materials Group at the University of Agder, Norway

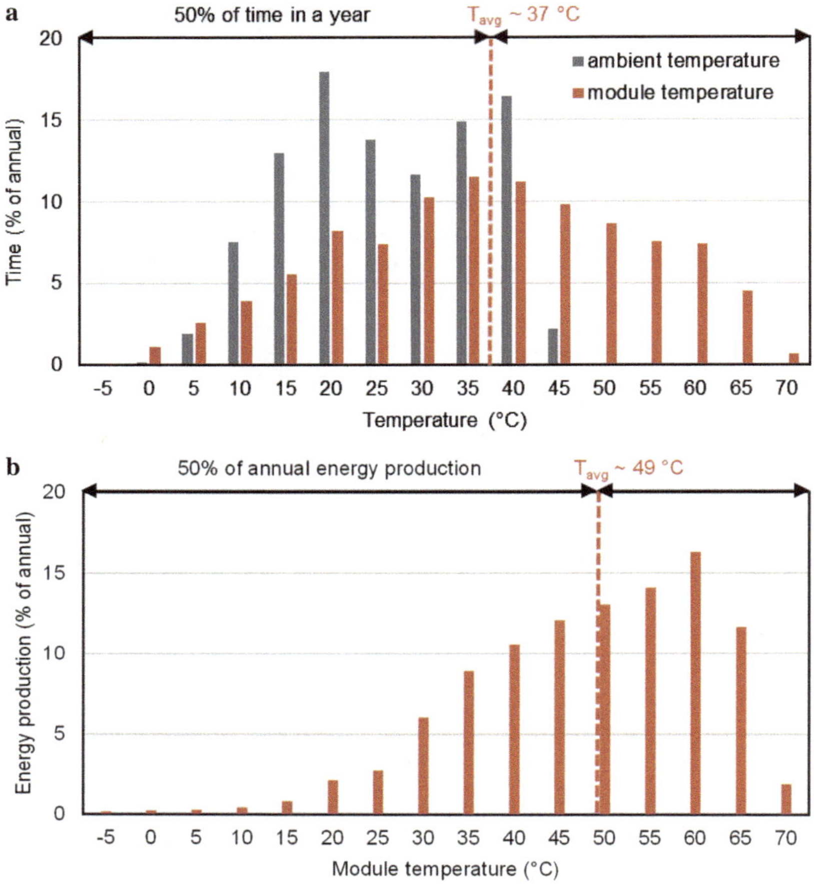

Fig. 1.3 Temperature distribution under real operating conditions of CdTe modules located in a power plant in the US desert southwest over: **a** time in a year; **b** annual energy production. Figure created with data from Strevel et al. (2012)

and Gonzalez 1980). Figure 1.5 shows field data of the power output versus module temperature and irradiance of a pc-Si[6] module installed in Italy (Moser et al. 2013). As for any PV installation, the operating conditions cover a large range of temperatures and solar irradiances. In this case, the standard test conditions are hardly ever met. The NOCT is certainly more representative of the device operating

[6]pc-Si stands for polycrystalline silicon. It is often used as a synonym of multi-crystalline silicon (mc-Si) but these denominations refer to materials with different grain sizes according to the terminology defined by Basore (1994). Grain sizes >10 cm: c-Si (monocrystalline silicon); 1 mm–10 cm: mc-Si (multicrystalline silicon); 1 µm–1 mm: pc-Si; <1 µm: µc-Si (microcrystalline silicon).

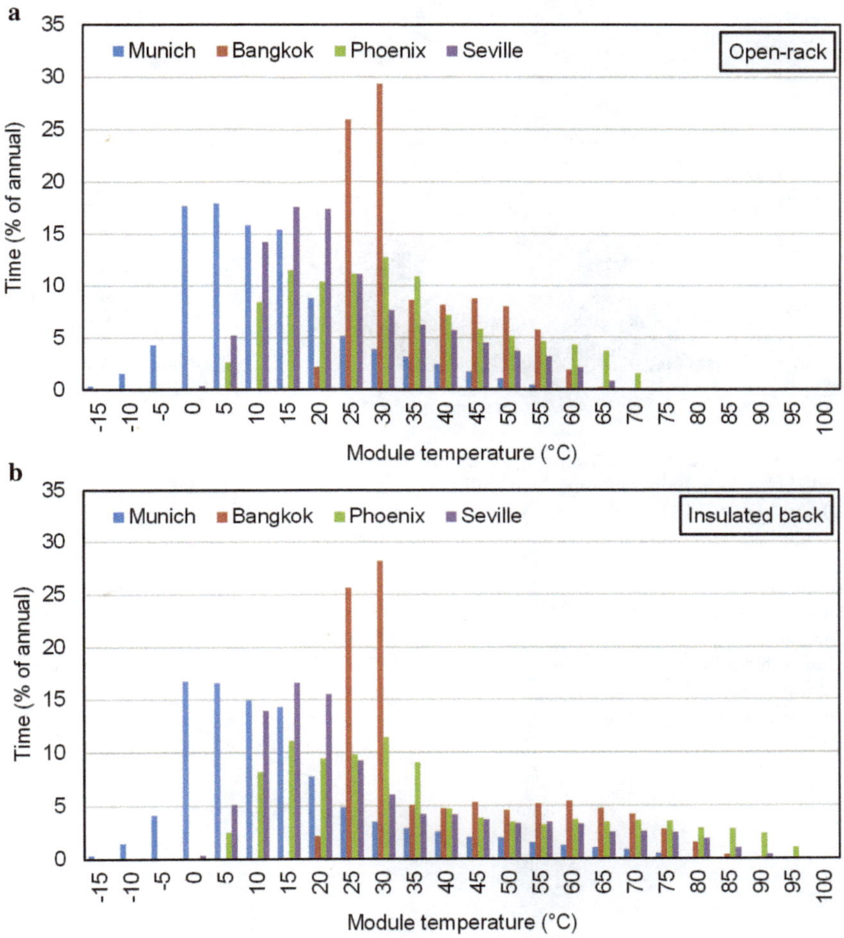

Fig. 1.4 Temperature distributions of PV modules in several locations for two different mounting configurations: **a** Open-rack; **b** Insulated back. Figure created with data provided by Sarah Kurtz, NREL

temperature but does not represent the large temperature variations that this module endures. Furthermore, as illustrated in Fig. 1.6, the NOCT gives only an idea of the quality of the thermal design of the module for a specific couple of solar irradiance and mounting configuration, and thus again cannot alone represent the variability of these two factors.

Nevertheless, there exists an abundant number of correlations aimed at predicting the cell operating temperature in a broader range of operating conditions than those corresponding to the NOCT. For example, a simple way to account for the dependence on irradiance (G) in the determination of the cell temperature (T_c) is

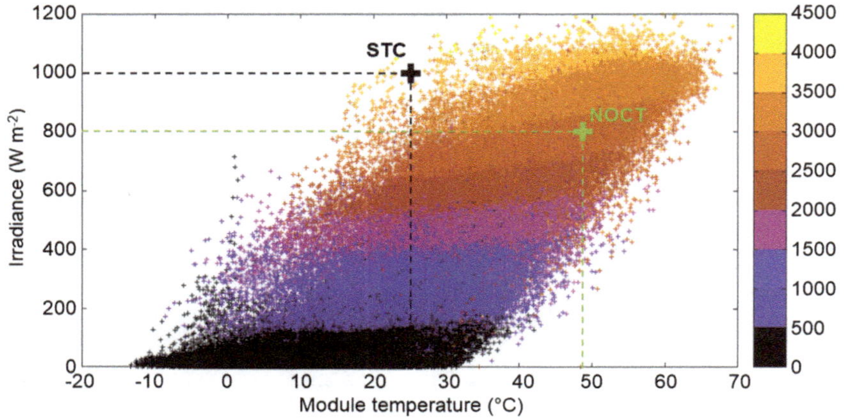

Fig. 1.5 Field data of the power output versus module temperature and irradiance for a pc-Si module installed in Italy. Figure created with data from the ABD-plant in Bolzano (Italy) provided by David Moser, EURAC Research

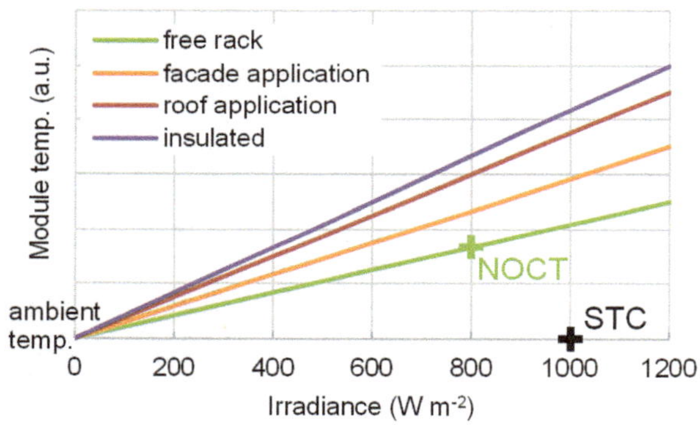

Fig. 1.6 Temperature difference between module and ambient as a function of irradiance for different mounting configurations. Figure created with data from Bloem (2008)

to assume a linear variation of the form $T_c = T_{amb} + G/800\ (NOCT\text{-}20)$, where T_{amb} is the ambient temperature. The review article (Skoplaki and Palyvos 2009a) provides several lists of correlations for the PV module operating temperature that are either explicit or implicit. In the latter case, some parameters in the formulas depend on temperature and thus an iterative procedure is required. The parameters used in these correlations include the ambient temperature, sometimes the back-side cell temperature, the heat transfer coefficient that scales the ability of the module to dissipate heat by convection, certain parameters in a reference set of conditions (STC or those for NOCT), absorptivity of the covering glass, absorptivity of the cell, emissivity of the module to account for dissipation by radiation with the cold outer space, etc. The same review article (Skoplaki and Palyvos 2009a) gives an overview of how the cell temperature is handled in software codes that predict PV array performances. For example, in the version 5.3.0 of the U.S. National

Renewable Energy Laboratory (NREL) PVWatts online program (Dobos 2014), one input parameter for calculating the cell temperature is the "installed" nominal operating cell temperature (INOCT). This parameter can be estimated from the NOCT and the mounting configuration or from field measurement data (Fuentes 1987).

Predicting the operating temperature of PV installations otherwise consists in establishing and solving the thermal balance equation for the devices, either in steady-state or time-varying operating conditions (for solar cells, the local climate reflected by the solar illumination, the ambient temperature and wind conditions). Usual models for the energy and thermal balances (e.g. Krauter and Hanitsch 1996; Jones and Underwood 2001; Silva et al. 2010) rely on the basic principles depicted in Fig. 1.7. Operating conditions are defined by the illumination (P_{inc}), the ambient temperature (T_{amb}), the temperature of the front and back at which the device exchanges thermal power usually by convection and/or conduction with its surroundings (T_{front} and T_{back} respectively), the environment temperature (T_e) (sometimes also the ground temperature T_{ground}) and possibly the cold outer space temperature (T_{sky}, see Sect. 1.3.4) with which the device exchanges thermal power by radiation. It is reasonably common to assume that the heat power that the device has to dissipate (Q_{heat}) is simply the radiation absorbed by the device (equal to the incident radiation power P_{inc} minus the reflected P_{refl} and transmitted radiation powers P_{trans}), which is not converted into electrical power (P_{PV}). As the electrical power output depends on the cell temperature (T_c) via the temperature coefficient (see Sect. 1.1.1), the energy and thermal balance equations are coupled. While this problem may look straightforward to solve, in fact, a large number of parameters have to be known in addition of those defining the operating conditions. These are the spectral radiative properties of the different layers constituting the device, both for the illumination wavelength range and in the infrared where thermal radiation is exchanged with the environment and the cold outer space. Their accurate determination, either from modeling or measurements, is important because the reflected radiation power has an impact on the amount of heat to dissipate and thus on the equilibrium temperature (see for example Silva et al. 2010; Gentle and Smith 2016). A temperature gradient across the device can be calculated by solving the

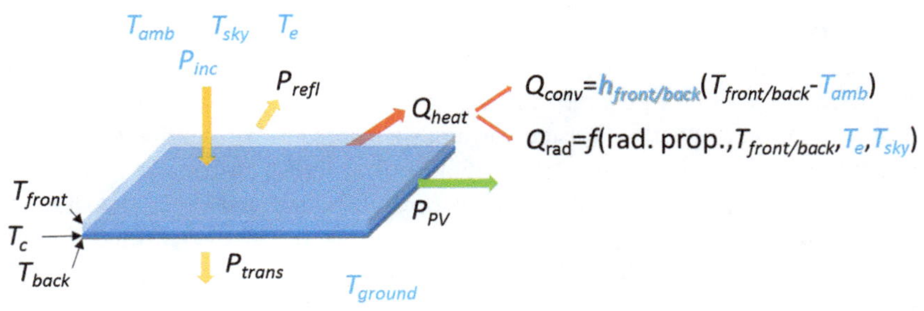

Fig. 1.7 Principles and parameters used in models for the energy and thermal balances of photovoltaic devices

conduction heat transfer equations, although it may turn out to be negligible (e.g. Notton et al. 2005; Hara et al. 2016; Weiss et al. 2016). Depending on the mounting configuration and operating conditions, modeling the heat transfer coefficient (h) at the front and the back of the device for free or forced convection can be challenging [see for example a review of correlations for the case of building integrated installations in Palyvos (2008)]. Rapid variations of irradiance require a transient modeling of the thermal balance of the device that accounts for the thermal inertia of the device (see e.g. Armstrong and Hurley 2010; Kant et al. 2016). It is also necessary to take into account the temporal variations of the absorbance of the device (e.g. Hoang et al. 2014). It is worth mentioning that simulation tools such as SILVACO Atlas (Silvaco 2016) and COMSOL-PV (Nardone and Green 2015) have thermal modules for calculating the equilibrium temperature of PV devices.

Accurately modeling air flow and convective heat transfer at the front and back surfaces of modules is a challenging task. For example, an assessment of relatively simple models that take into account the effect of wind to predict the module temperature indicates that, in the case a ground-mounted PV plant, large discrepancies can be found among the selected correlations and their application to different PV technologies (Schwingshackl et al. 2013). This is why there is a growing number of computational fluid dynamics (CFD) analyses aimed at estimating the wind loads for ground-mounted arrays of solar panels (e.g. Shademan and Hangan 2009; Aly 2016; Jubayer and Hangan 2016). Similarly, CFD studies are being conducted for optimizing the cooling of modules in Building Integrated Photovoltaic systems (BIPV, see Sect. 1.3.3) (see e.g. Brinkworth et al. 1997; Bloem 2008; Lau et al. 2012).

Alternatively to the predicting models, there exist semi-empirical models established on the prior building of experimental databases, which require the system to be installed and equipped with sensors. For example, (King et al. 2004) reports on a correlation that involves the incident irradiation (P_{inc}), wind speed measured at ten meters above the ground (WS), ambient temperature (T_{amb}) and two empirical parameters (a and b) used to estimate the back-surface module temperature ($T_{back} = G\ \exp(a + b\ WS) + T_{amb}$). The empirically determined parameters can be adjusted to specific locations and wind speed measurements. Applied to flat-plate modules mounted on open racks or with insulated back surfaces, and to concentrator modules with finned heat sinks, this model is reported to give module operating temperatures with an uncertainty of about ±5 °C, and subsequently of 3% on the power output (King et al. 2004). The difference between the cell and the back-surface module temperatures is measured for a standard irradiance (1000 W m^{-2}) and linearly adjusted for other irradiances. This difference depends on module type and mounting configuration. For example, in King et al. (2004), it is stated to be zero and three degrees for back-insulated and open-rack mounted modules, respectively. Temperature can be inhomogeneous over a module array, usually by less than 5 °C according to King et al. (2004). This raises the question of how to determine accurately an average temperature and subsequently temperature coefficients for an entire array (Guay et al. 2016).

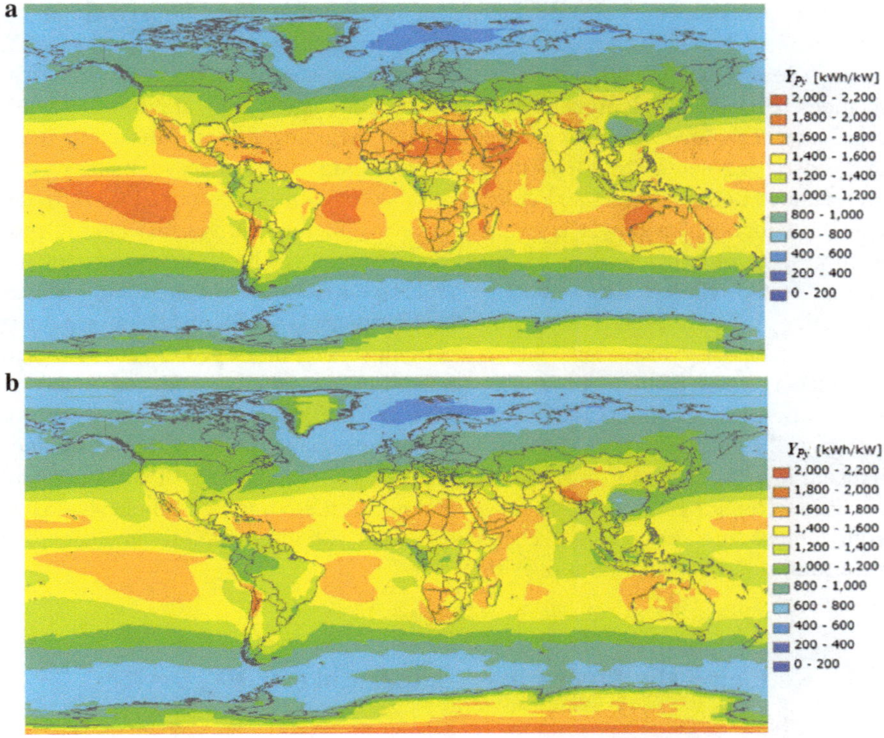

Fig. 1.8 Potential world map of energy yield for c-Si modules: **a** without temperature effects; **b** with temperature effects. Reprinted with permission from Kawajiri et al. (2011). Copyright 2016 American Chemical Society

By accounting for local climate parameters, namely the irradiation and the ambient temperature, the effect of temperature on the potential of PV generation all over the world can be estimated (Kawajiri et al. 2011). The world maps of potential energy generation by c-Si PV modules with or without taking into account temperature effects are shown in Fig. 1.8. The first map, Fig. 1.8a, where the effect of temperature is not taken into account, is quite similar to a map of annual solar irradiance. The second map, Fig. 1.8b, where the negative impact of temperature on PV conversion is considered, shows some modifications in the regions of highest PV production potential. On the one hand, the PV potential is reduced in warm locations and, on the other hand, high mountain ranges where illumination is important and temperature is low are highlighted. These observations promote the idea of designing and selecting modules according to the local climate. For example, modules with temperature coefficients smaller than that of c-Si would be more efficient in the warmer regions. Prediction of PV energy yield as a function of temperature is the topic of a great number of analyses. The review from Skoplaki and Palyvos (2009b) reports that almost all listed correlations for the power output and the conversion efficiency are linear with device temperature. Some correlations are extended with climate parameters (such as irradiation, wind speed, etc.), while others use the NOCT instead of the device temperature. Some semi-empirical methods consist in adopting a model with parameters that are inferred with regression techniques from experimental data, including the module temperature (e.g. Rosell and Ibanez 2006). Methods for predicting annual energy production can be separated between direct and indirect approaches (Rus-Casas et al. 2014). In the latter category, either power is evaluated directly before deriving an energy production [methods reviewed in Skoplaki and Palyvos (2009b)], or models for the current-voltage characteristics of the generator are used prior to determining power. Predicting, measuring and interpreting the energy yield of PV installations is certainly not an easy task. This fact is well illustrated in Hegedus (2013) which provides a review of the energy yield (defined by the ratio of produced energy in kWh to power rated in STC in kW) measured in various locations in the world for different PV technologies. Variability in the observed trends and uncertainties indicate that the assessment of performances of PV installations in real operating conditions requires an accurate understanding of the combined efficiency losses with temperature (temperature coefficient) and with illumination (intensity and spectrum) and their evolution over time.

1.3 Reducing the Operating Temperature of PV Devices

Since it is long known that temperature reduces the performances of PV devices, a large amount of research has been aimed at minimizing the operating temperature of PV modules. Most of the existing strategies consist in evacuating the heat generated in the devices. Some works study the use of cooling systems (passive or active) while other discuss the design of components (e.g. frame, protective sheets, etc.) with high thermal conductivities. While it is not common to use cooling systems for standard PV applications due to the additional cost, specific applications such as concentrated PV, building integrated PV (BIPV) or space missions are driving the development of cooling strategies (e.g. Herschitz and Bogorad 1994; Royne et al. 2005; Chandrasekar et al. 2015).

Section 1.3.1 describes some of the various existing strategies used for cooling PV systems. Section 1.3.2 presents the concept of hybrid photovoltaic-thermal collectors which makes use of the heat generated in the PV modules. Section 1.3.3 describes opportunities of thermal designs in specific applications (BIPV and floating PV installations). Section 1.3.4 discusses the potential of radiative cooling for solar cells with a special focus on the opportunity provided by the atmospheric spectral windows to exchange radiation with cold outer space. An emergent approach to minimize the operating temperatures without adding to the system cost and complexity consists in minimizing by design the heat generated within the devices. Section 1.3.5 reviews the design strategies that take the thermal behavior of PV devices into account to reduce the temperature-induced losses.

1.3.1 Common Cooling Strategies

There exists a myriad of different cooling systems. The cooling media can be air with a fan to force a flow around the module or water pumped through a heat exchanger on the rear of the module. It is also suggested to cool PV modules via water flowing on their front surfaces (Krauter 2004; Odeh and Behnia 2009). A water film provides the supplementary benefits of cleaning the surface of the modules and serving as an anti-reflection layer. Using this technique, and taking into account the power consumed by the pumping system, a net gain of about 8% is claimed in Krauter (2004). Different gains in energy yield and reductions of operating temperature are reported in the literature for various cooling methods [see the review in Chandrasekar et al. (2015)]. In any case, the energetic and economic viability of the cooling system has to be properly assessed. When water is used, the issue of water consumption has to be addressed [in Krauter (2004), the evaporation rate is estimated to about 1 l h^{-1} m^{-2}].

The need for cooling is particularly important in concentrated PV applications. Indeed, as the illumination increases, so does the heat generated within the device which needs to be evacuated. The various solutions suggested for cooling PV cells under concentration are reviewed in Royne et al. (2005). Figure 1.9 shows the range of thermal resistances (i.e. the inverse of heat transfer coefficients) that can be achieved via different cooling system technologies.

Until recently, cooling of PV systems was used mainly in specific applications such as concentrated PV (where the heat load is particularly important), BIPV [where the operating temperature can rise above 90 °C (Virtuani et al. 2010) when convective heat transfer at the back of the modules is limited], or space missions (that can be subject to very exotic operating conditions). However, the PV industry

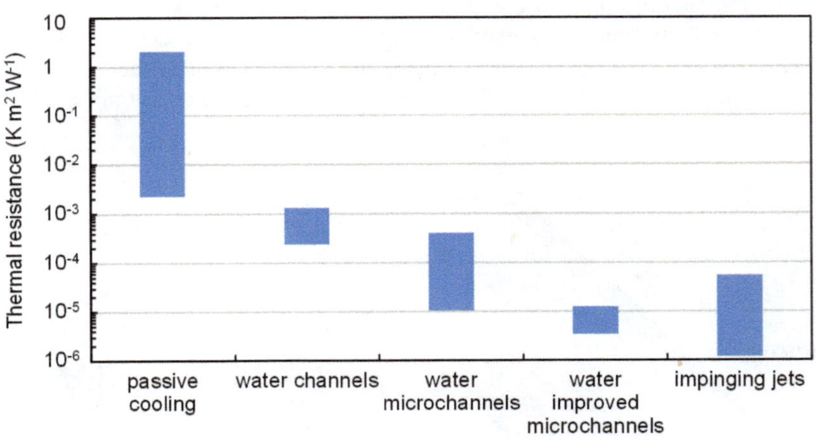

Fig. 1.9 Range of thermal resistances reported for different cooling technologies. Figure created with data from Royne et al. (2005)

is starting to realize fully the impact of operating temperature on the energy yields of PV installations and startup companies that propose intelligent cooling solutions are emerging. For example, Sunibrain commercializes a cooling and cleaning system using sprinklers that can be added on existing PV installations (Cristi et al. 2016). The system collects rainwater falling on the modules into a storage tank and uses it to cool the modules when it is most needed. Despite the fact that the water is recycled, water availability is limited, especially in summer. This is why water management is optimized by a specific algorithm associated with an Artificial Intelligence (AI) platform that has access to several local data (PV module production and temperature, water temperature and availability, etc.) and online weather forecasts (Macq et al. 2016).

1.3.2 Hybrid Photovoltaic-Thermal Solutions

The current average efficiency of commercial wafer-based silicon modules is about 17% (Fraunhofer 2016). This means that, assuming 5% of radiative losses, about 78% of the total incident solar energy is transformed into heat within the modules. The idea motivating hybrid photovoltaic-thermal (PV-T) solutions is to take advantage of this heat source. By combining a photovoltaic unit and a solar thermal collector, PV-T systems can use a larger fraction of the solar energy and thus have the potential to reduce the use of material and required space (Good 2016). The different technologies and designs of PV-T systems are reviewed in several articles (e.g. Chow 2010). To optimize the overall benefit, the whole PV-T system has to be designed as a function of the desired thermal to electrical yield ratio which drives the choice of the collector type, PV surface, working temperature, coolant medium (usually air or water), fanning or pumping system, etc.

While the dual use of solar energy enabled by PV-T systems may be interesting in certain configurations, it is necessary to consider the increased cost and complexity of the system (and the energy potentially required to drive the cooling fluid) compared to a standard PV installation. Figure 1.10 illustrates a domestic hybrid

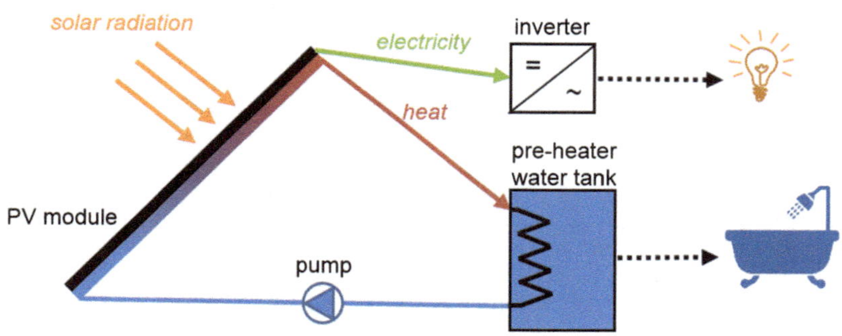

Fig. 1.10 Schematic illustration of a domestic hybrid photovoltaic-thermal (PV-T) installation

PV-T installation where the heat extracted from the PV modules is used to pre-heat a domestic hot water tank. In this kind of installation, there is a trade-off between the electrical efficiency and the temperature of the thermal source: the temperature of the cooling fluid should be low to keep the PV module cool but also needs to be high enough to pre-heat efficiently the domestic water above room temperature.

One interesting feature of hybrid PV-T systems is their flexibility: the working conditions (temperature, coolant flow rate, etc.) can be modified at any time to adapt the electricity-to-thermal performance ratio to the energy needs of the building.

1.3.3 Building Integrated PV and Floating PV

Certain applications offer interesting design opportunities to mitigate the thermal losses of photovoltaic modules. In building integrated photovoltaics (BIPV), the photovoltaic converters replace partially or completely a traditional building element (rooftop, façade, etc.). In these BIPV applications, the PV devices may have other functionalities/requirements than the generation of electricity such as soundproofing, water sealing, thermal insulating, etc. Certain applications require more or less transparent devices to let some of the light enter into the building. This can be achieved using certain PV technologies (Heinstein et al. 2013). Integrating PV in the building envelope also creates some constraints on aesthetics: devices can be specifically designed, sometimes at the expense of efficiency, to have a desired appearance [e.g. white modules, (Escarré et al. 2015)].

When the PV modules are directly attached to the building envelope, their temperature can reach over 90 °C (Virtuani et al. 2010). This drastically reduces the photovoltaic performances and potentially increases the heat load that need to be dissipated by the building air conditioning system during summer. Because of this well-known thermal issue, BIPV designs have evolved into PV units that are detached from the building (such as double skin facades) and use heat transfer mechanisms to remove the heat from the modules and eventually reuse it to pre-heat air supplied to the building interior (BIPV-T designs). The medium used for heat transfer can be air, water, other working fluids, thermoelectric materials, phase-change materials, etc. (Lai and Hokoi 2014).

Double skin façades or roofs offer interesting opportunities for integrating PV into buildings (see Fig. 1.11). A double skin façade is used as a thermal buffer layer, sound protection layer, and eventually as a solar preheater of ventilation air (Agathokleous and Kalogirou 2016). When used in combination with PV panels, the double skin façade also provides an efficient passive cooling method.

Another example of a specific configuration which allows for an efficient passive cooling of PV modules is floating photovoltaics [see (Trapani and Redon Santafé 2014) for a review of floating photovoltaic projects between 2007 and 2013]. Floating photovoltaics, sometimes called floatovoltaics, refers to PV systems

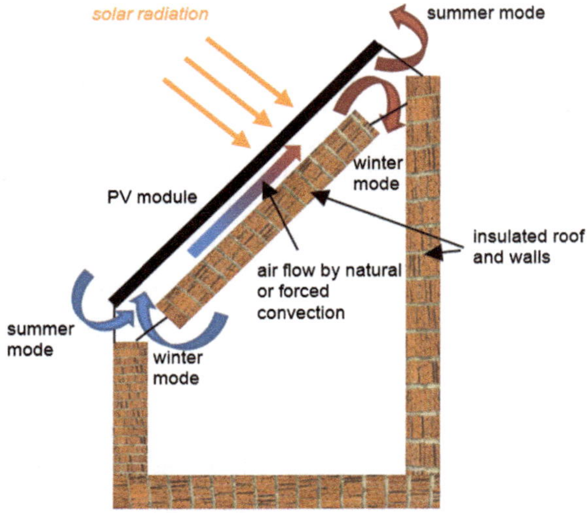

Fig. 1.11 Schematic illustration of the operating principle of a double skin roof

designed to float on water (see Fig. 1.12a). This concept has several advantages: it makes use of surfaces such as dam reservoirs that are otherwise unoccupied; it can also reduce water evaporation from the aforementioned water reservoirs and reduce algae growth; the floating PV devices operate in a low dust environment and can be easily cleaned (using sprinklers for example); in dam reservoir-based hydro power plants, the energy generated by the floating PV plant can be efficiently stored by pumping water from a lower reservoir; finally the PV panels benefit from the natural evaporative cooling of the water body (pond, lake, dam reservoir, etc.). This cooling effect would result in gains in energy yield ranging from 20 to 70% according to different developers (Trapani and Redon Santafé 2014). In order to maximize cooling, an approach quite similar to floatovoltaics is developed by a research group from Italy where the PV modules are completely submerged at a water depth of a few centimeters (Tina et al. 2012). According to Tina et al. (2012), the loss due to absorption of light by water can be more than compensated by lower operating temperature of the submerged PV devices. Despite several advantages, floating and submerged PV installations suffer from several technical complexities (such as water infiltration, corrosion and maintenance issues) which make guaranteeing a 20 year lifetime challenging.

Another original configuration enabling a different kind of passive cooling is the installation of a PV plant in synergy with a green roof (see Fig. 1.12b): modules provide shadowing while the plants contribute to lowering ambient temperature (Witmer 2010).

Fig. 1.12 Photographs of: **a** floating photovoltaic modules © Ciel & Terre International, French manufacturer of the patented floating solar solution Hydrelio®. Keens Farm; **b** photovoltaic modules on a green roof © Optigrün international AG, Gunter Mann, 2015

1.3.4 Radiative Cooling

Although radiative cooling is an important process in determining the temperature of standard photovoltaic modules, this cooling is often envisaged and modeled in terms of radiative transfer between the module and both ground and sky, with these two media at temperatures not too different from ambient.

However, recent work points out that the atmospheric spectral window between 8 and 13 micron wavelength allows direct communication with cold space, where the background temperature is only 3 K (Raman et al. 2014). The possibilities opened up by this consideration are highlighted by the design of surfaces that cool to temperatures below ambient even when exposed to direct sunlight (Raman et al. 2014; Gentle and Smith 2015).

The potential of harnessing the "coldness of the universe" in this way to improve photovoltaic module performance, perhaps even allowing modules to operate below ambient temperature, can be evaluated by calculating the hemispheric emission through a perfectly transparent window over the above wavelength range. This is calculated as 133 W m^{-2} at 20 °C compared to the value of 40 W m^{-2} that is experimentally demonstrated in Raman et al. (2014). Thus, if the theoretical limit were achieved, a PV cell with a heat source smaller than 133 W m^{-2} would be able to operate at a sub-ambient temperature. Considering the heat sources depicted in Fig. 3.8a, this could be achieved by a PV device operating in the radiative limit with a perfect sub-bandgap photon reflector and with a bandgap larger than ~1. 6 eV. In this configuration, conversion efficiencies up to 30% could be reached in real operating conditions (see also Safi and Munday 2015). A non-ideal cell would have a larger heat source and thus would need to have a larger bandgap and a corresponding smaller efficiency to operate at temperatures below ambient (see Fig. 3.8b). This cooling effect improves with increasing cell temperature (for example, it is calculated as 228 W m^{-2} at 57 °C, a temperature which is in the usual range of operating temperatures of PV devices).

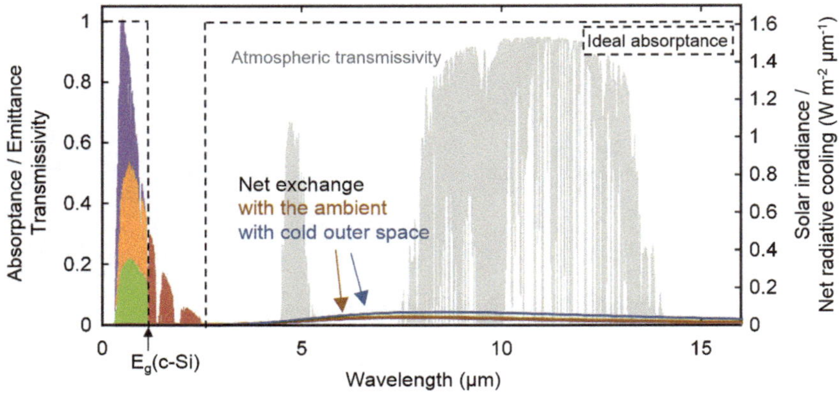

Fig. 1.13 Ideal absorptance profile to contribute to minimizing the operating temperature of a c-Si photovoltaic device (*dashed line*). The AM1.5 solar irradiance is split in *purple*, *red*, *orange* and *green* areas which correspond to the heat generated by thermalization, the heat source avoided by the reflection of sub-bandgap photons, the other heat sources associated with PV conversion, and the output power, respectively. The net exchanges with the ambient and cold outer space are shown in *brown* and *blue*. The exchange with cold outer space is only possible through the atmospheric windows shown in *grey*

As mentioned above, it is ideal in terms of thermal behavior that a solar cell reflects all sub-bandgap photons to prevent the parasitic absorption of these photons useless for photovoltaic conversion (see Sect. 3.5.1). Figure 1.13 illustrates the ideal absorptance profile: a device fully absorbing above the bandgap, then fully reflective to circa 2.5 microns, then again fully absorbing at longer wavelengths. Note that the wavelength above which absorption should be maximum is the wavelength at which the net radiative cooling between the cell and its surroundings is larger than the heating power of the sub-bandgap solar photons. For standard terrestrial conditions (AM1.5 spectrum and an environment at 25 °C), this wavelength is about 2.5 microns—where solar irradiance becomes negligible (see Fig. 2.10)—but this threshold can be different in specific configurations such as TPV applications (see Sect. 3.5.1).

Standard modules, not reflecting the sub-bandgap photons, are constrained to operating temperatures linked to sky, ambient and ground temperatures. Radiative cooling of current PV modules is already close to its maximum since low-iron glass coversheets are absorbing beyond 4 micron (Rubin 1985), although improvements have been suggested using micron-sized, pyramidal-texturing of top glass surfaces (Zhu et al. 2014). Interestingly, these features have been used in high-performance modules, but rather to improve reflection and angular response (Zhao et al. 1999). The accidentally good infrared emission properties of low iron glass coversheets limit the gains that can be achieved by design modifications that attempt to exploit this effect. However, some improvements could be obtained by designing a coating whose angular absorptance profile in the infrared remains high at very high angles (Gentle and Smith 2016).

1.3.5 Thermal Design of PV Cells and Modules

In order to optimize their field performances, it is possible to take into account the thermal behavior (heat generation and dissipation) of PV cells and modules in their design.

Temperature-induced losses are especially critical for PV devices designed for space applications. There is no convection in space so the heat generated in the device can only be dissipated by thermal radiation. As mentioned above, radiative cooling can be somewhat increased by improving the device emissivity in the appropriate wavelength range (Zhu et al. 2014). A complementary approach consists in reducing the heat source in the solar cell. This can be done using multilayer coatings that reflect as much of the useless incident light as possible, i.e. in spectral regions beyond the limits of the cell response (Beauchamp et al. 1993; Beauchamp and Tuttle-Hart 1995). The corresponding reduction in cell absorptance leads to a reduction of the heat generated in the device and ultimately in a reduction of its operating temperature. This concept is also investigated for terrestrial high-concentrating photovoltaic systems (e.g. Maghanga et al. 2011; Micheli et al. 2012). For some specific space applications, the thermal management of the PV system is done at the expense of its performances. For example, tilting the arrays away from the Sun is considered to prevent them from overheating during near-sun missions (Landis et al. 2004).

A simple design strategy that takes into account the operating temperature of standard terrestrial PV modules consists in tilting their azimuth slightly towards the east (Krauter and Hanitsch 1996). The tilt has to be eastward because the ambient temperature generally lags behind solar irradiance intensity as illustrated in Fig. 1.14. This strategy enables to lower the reflection in the morning when ambient temperatures are lower. By using this method, a gain in energy output of 0.4% is claimed in Krauter and Hanitsch (1996). Another original idea is suggested in Krauter and Hanitsch (1996): it consists in adding a small water tank to the back of a PV module in order to increase its heat capacity so that its maximum temperature is lowered and is reached after the time of maximum incident irradiation. Figure 1.15 illustrates that this "heat capacity" concept provides an efficient mean to reduce the operating temperature of the PV module during the hours of maximum irradiation (and thus of large power production). However, one important drawback is the significant weight added to the PV unit[7] that makes installation (especially roof mounting) more difficult.

Chintapalli et al. (2010) propose a modified design of PV modules that aims at maximizing heat dissipation. Based on an earlier study (Lee et al. 2008), they suggest to use a modified EVA composite encapsulant with a thermal conductivity larger than that of the classic EVA encapsulant (and still sufficiently electrically insulating). They also propose to replace the polyvinyl fluoride backing of traditional modules by an aluminum sheet with fins in order to enhance the heat transfer

[7]The prototype modules in Krauter (2004) weight 200 kg.

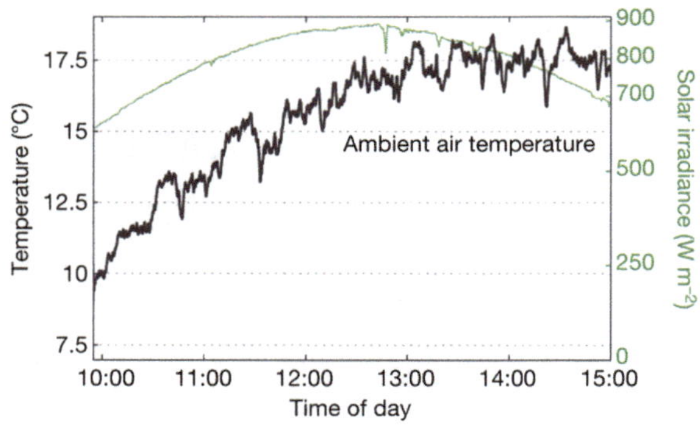

Fig. 1.14 Hourly variations of solar irradiance and ambient temperature. The shift between the maximum of solar irradiance and that of ambient temperature is visible. Figure from Raman et al. (2014)

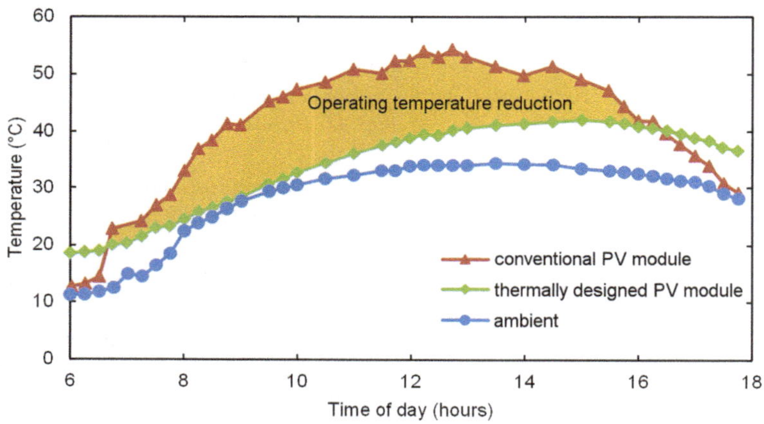

Fig. 1.15 Hourly variations of ambient, conventional PV module and thermally and optically optimized PV module temperatures. The temperature lag and the maximum temperature reduction of the thermally and optically optimized PV module are apparent. Figure created with data from Krauter et al. (1996)

coefficient at the rear of the module. Finally, they suggest replacing flat glass with textured glass to increase heat transfer at the front of the module (because the surface area is increased by about 30%). The experimental results (Chintapalli et al. 2010) show that the aluminum finned substrate is the most efficient among the different solutions in terms of heat dissipation. It reduces the operating temperature of the module by about 10 °C. Other studies investigating the benefit of using aluminum finned substrates report gains in operating temperature of about 14 °C (Ross and Gonzalez 1980) and 17 °C (Cuce et al. 2011) under an illumination of 800 W m^{-2}. While the thermal advantage of this solution is demonstrated, the question of its economic viability remains.

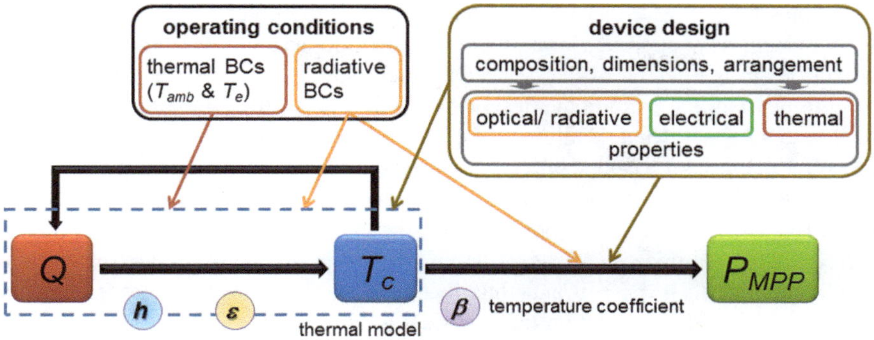

Fig. 1.16 Schematic of the thermal behavior of a photovoltaic device and its dependence on operating conditions and device design

The different examples presented above illustrate that it is possible to increase the energy yield of photovoltaic systems by taking into account their thermal behavior in their design in order to minimize their operating temperature. Most of the existing strategies aim at maximizing heat dissipation and some original strategies target the minimization of heat generation. In order to investigate all the possible options to reduce the temperature-induced losses, it is necessary to use a comprehensive model of the thermal behavior of PV devices. A simplified schematic of such a model is shown in Fig. 1.16. It illustrates that optical, electrical and thermal mechanisms are deeply intertwined. Thus, taking into account a thermal criterion (e.g. minimizing the device operating temperature) modifies the optimal values of all the design parameters of a PV device (e.g. semiconductor material, layers' thicknesses and doping, architecture, contacts, etc.). Additionally, these optimal values are function of the device operating conditions which depend on the mounting configuration and the climate in which the device is to be installed in.

In Fig. 1.16, the symbol Q corresponds to the heat generated within the device. The cell temperature (T_c) is driven by the heat source (Q) and the thermal and radiative Boundary Conditions (BCs). These boundary conditions include in particular the ambient temperature (T_{amb}), the intensity and spectrum of the incident radiation, the heat transfer coefficient (h) which takes into account conduction and convection, and the emittance (ε) which drives the radiative exchanges between the system and its surroundings. The cell temperature impacts the output electrical power via the so-called temperature coefficient (β). Chapter 2 gives a detailed description of the physics involved in the temperature coefficients of PV devices and provides some guidelines for tuning them. Figure 1.16 also illustrates that the heat generated within a PV device is driven by the operating conditions and the device design. For example, the thermalization of the photogenerated electrical charges depends on the spectrum and intensity of the incident radiation and on the interband absorption coefficient and thickness of the absorbing layer. Chapter 3 provides a comprehensive and general thermal model for PV devices which describes all the heat generation and dissipation mechanisms. It also presents guidelines for going further than the standard test conditions in the optimization of PV devices. The approach consists in taking into account the actual operating conditions in the design of the devices by including thermal criteria. One example of optimization that include the design rule "minimizing the internal heat load" can be found in Dupré and Vaillon (2014). In this work, the anti-reflection coating (ARC) thickness of a c-Si solar cell is optimized for different operating conditions. The results show that when the heat transfer coefficient is limited (as is the case in certain building integrated applications), the thickness of the ARC should be increased to minimize the absorption of high energy photons which contribute significantly to heat the cell and thus increase the operating temperature. Several other examples of optimization that consider a thermal criterion are given in Chap. 3. It is argued that optimizing the whole set of design parameters of PV (or TPV) devices, as a function of their operating conditions, will enable significant improvements in the energy yields of future PV installations.

References

Agathokleous RA, Kalogirou SA (2016) Double skin facades (DSF) and building integrated photovoltaics (BIPV): a review of configurations and heat transfer characteristics. Renew Energy 89:743–756. Doi:10.1016/j.renene.2015.12.043

Aly AM (2016) On the evaluation of wind loads on solar panels: the scale issue. Sol Energy 135:423–434. Doi:10.1016/j.solener.2016.06.018

Armstrong S, Hurley WG (2010) A thermal model for photovoltaic panels under varying atmospheric conditions. Appl Therm Eng 30:1488–1495. Doi:10.1016/j.applthermaleng.2010.03.012

Basore P (1994) Defining terms for crystalline silicon solar cells. Prog Photovoltaics Res Appl 2:177–179

Beauchamp WT, Tuttle-Hart T (1995) Patent n° US5449413 A—"UV/IR reflecting solar cell cover"

Beauchamp WT, Tuttle-Hart T, Sanders ML (1993) Blue/red reflecting solar cell covers for GaAs cells. In: Proceedings of the 23rd IEEE photovoltaic specialist conference. pp 1487–1490

Bloem JJ (2008) Evaluation of a PV-integrated building application in a well-controlled outdoor test environment. Build Environ 43:205–216. Doi:10.1016/j.buildenv.2006.10.041

Branz HM (2003) The hydrogen collision model of metastability after 5 years: experimental tests and theoretical extensions. Sol Energy Mater Sol Cells 78:425–445. Doi:10.1016/S0927-0248(02)00446-4

Brinkworth BJ, Cross BM, Marshall RH, Yang H (1997) Thermal regulation of photovoltaic cladding consequences. Sol Energy 61:169–178

Carlson DE, Lin G, Ganguly G (2000) Temperature dependence of amorphous silicon solar cell PV parameters. In: Proceedings of the 28th IEEE photovoltaic specialist conference

Chandrasekar M, Rajkumar S, Valavan D (2015) A review on the thermal regulation techniques for non integrated flat PV modules mounted on building top. Energy Build 86:692–697. Doi:10.1016/j.enbuild.2014.10.071

Chintapalli M, Diskin M, Guha I (2010) Improving solar cell efficiency: a cooler approach. http://web.mit.edu/3.042/team5_10/Poster.html. Accessed 22 Aug 2016

Chow TT (2010) A review on photovoltaic/thermal hybrid solar technology. Appl Energy 87:365–379. Doi:10.1016/j.apenergy.2009.06.037

Cristi N, Macq A, Martin-Carron L (2016) Evaporating pure rainwater to increase the yield of commercial-size PV arrays. In: Proceedings of the 32nd European photovoltaic solar energy conference

Cuce E, Bali T, Sekucoglu SA (2011) Effects of passive cooling on performance of silicon photovoltaic cells. Int J Low-Carbon Technol 6:299–308. Doi:10.1093/ijlct/ctr018

Dobos AP (2014) PVWatts Version 5 Manual (NREL/TP-6A20-62641)

Dupré O, Vaillon R (2014) Optimizations of photovoltaic cells including the minimization of internal heat sources. J Renew Sustain Energy 6:11201. Doi:10.1063/1.4828367

Dupré O, Vaillon R, Green MA (2015) Experimental assessment of temperature coefficient theories for silicon solar cells. IEEE J Photovoltaics 1–5. Doi:10.1109/JPHOTOV.2015.2489864

Escarré J, Li H, Sansonnens L et al (2015) When PV modules are becoming real building elements: white solar module, a revolution for BIPV. In: Proceedings of the 42nd IEEE photovoltaic specialist conference

First Solar (2014) First solar series 3 Black Plus™ PV Module

Fischer D, Bichsel F, de Bruijn S et al (2009) Positive effective temperature coefficient of power of +0.75%/°C in flexible a-Si modules in building integrated installations. In: Proceedings of the 24th European photovoltaic solar energy conference. pp 3505–3508

Fraunhofer (2016) Photovoltaics report

Fuentes MK (1987) A simplified thermal model for flat-plate photovoltaic arrays. Sandia report

Gentle AR, Smith GB (2016) Is enhanced radiative cooling of solar cell modules worth pursuing? Sol Energy Mater Sol Cells 150:39–42. Doi:10.1016/j.solmat.2016.01.039

Gentle AR, Smith GB (2015) A subambient open roof surface under the mid-summer sun. Adv Sci 2:1–4. Doi:10.1002/advs.201500119

Good C (2016) Environmental impact assessments of hybrid photovoltaic-thermal (PV/T) systems—a review. Renew Sustain Energy Rev 55:234–239. Doi:10.1016/j.rser.2015.10.156

Green MA (2003) General temperature dependence of solar cell performance and implications for device modelling. Prog Photovoltaics Res Appl 11:333–340. Doi:10.1002/pip.496

Guay NG, Hansen CW, Robinson CD et al (2016) Improving module temperature measurements using averaging resistive temperature devices. In: Proceedings of the 43rd IEEE photovoltaic specialist conference

Halsted RE (1957) Temperature consideration in solar battery development. J Appl Phys 28:1131. Doi:10.1063/1.1722592

Hara S, Kasu M, Matsui N (2016) Estimation method of solar cell temperature using meteorological data in mega solar power plant. IEEE J Photovoltaics 1–6. Doi:10.1109/JPHOTOV.2016.2589363

Hegedus S (2013) Review of photovoltaic module energy yield (kWh/kW): comparison of crystalline Si and thin film technologies. Wiley Interdisc Rev Energy Environ 2:218–233. Doi:10.1002/wene.61

Heinstein P, Ballif C, Perret-Aebi LE (2013) Building integrated photovoltaics (BIPV): review, potentials, barriers and myths. Green 3:125–156. Doi:10.1515/green-2013-0020

Herschitz R, Bogorad A (1994) Space environmental testing of blue red reflecting coverglasses for gallium arsenide and high efficiency silicon solar cells. In: Proceedings of the 1st IEEE world conference photovoltaic energy conversion (WCPEC). Doi:10.1109/WCPEC.1994.521657

Hoang P, Bourdin V, Liu Q et al (2014) Coupling optical and thermal models to accurately predict PV panel electricity production. Sol Energy Mater Sol Cells 125:325–338. Doi:10.1016/j.solmat.2013.11.032

Hoffmann S, Koehl M (2012) Effect of humidity and temperature on the potential-induced degradation. Prog Photovoltaics Res Appl. Doi:10.1002/pip

Jones AD, Underwood CP (2001) A thermal model for photovoltaic systems. Sol Energy 70:349–359. Doi:10.1016/S0038-092X(00)00149-3

Jubayer CM, Hangan H (2016) A numerical approach to the investigation of wind loading on an array of ground mounted solar photovoltaic (PV) panels. J Wind Eng Ind Aerodyn 153:60–70. Doi:10.1016/j.jweia.2016.03.009

Kant K, Shukla A, Sharma A, Biwole PH (2016) Thermal response of poly-crystalline silicon photovoltaic panels: numerical simulation and experimental study. Sol Energy 134:147–155. Doi:10.1016/j.solener.2016.05.002

Kawajiri K, Oozeki T, Genchi Y (2011) Effect of temperature on PV potential in the world. Environ Sci Technol 45:9030–9035. Doi:10.1021/es200635x

Kim TH, Park NC, Kim DH (2013) The effect of moisture on the degradation mechanism of multi-crystalline silicon photovoltaic module. Microelectron Reliab 53:1823–1827. Doi:10.1016/j.microrel.2013.07.047

King DL, Boyson WE, Kratochvill JA (2004) Photovoltaic Array Performance Model. Sandia report

Koehl M, Heck M, Wiesmeier S, Wirth J (2011) Modeling of the nominal operating cell temperature based on outdoor weathering. Sol Energy Mater Sol Cells 95:1638–1646. Doi:10.1016/j.solmat.2011.01.020

Kondo M, Nishio H, Kurata S et al (1997) Effective conversion efficiency enhancement of amorphous silicon modules by operation temperature elevation. Sol Energy Mater Sol Cells 49:1–6. Doi:10.1016/S0927-0248(97)00168-2

Krauter S (2004) Increased electrical yield via water flow over the front of photovoltaic panels. Sol Energy Mater Sol Cells 82:131–137. Doi:10.1016/j.solmat.2004.01.011

Krauter S, Hanitsch R (1996) Actual optical and thermal performance of PV-modules. Sol Energy Mater Sol Cells 41–42:557–574. Doi:10.1016/0927-0248(95)00143-3

Krauter S, Hanitsch R, Moreira L (1996) New optical and thermal enhanced PV-modules performing 12% better under true module rating conditions. In: Proceedings of the 25th IEEE photovoltaic specialist conference. Doi:10.1109/PVSC.1996.564377

Kurtz S, Whitfield K, Tamizhmani G et al (2011) Evaluation of high-temperature exposure of photovoltaic modules. Prog Photovoltaics Res Appl 954–965. Doi:10.1002/pip

Lai CM, Hokoi S (2014) Solar façades: a review. Build Environ 91:152–165. Doi:10.1016/j.buildenv.2015.01.007

Landis GA, Jenkins P, Scheimant D, Ryne R (2004) Extended temperature solar cell technology development. In: Proceedings of the 2nd international energy conversion engineering conference. pp 1–7

Lau GE, Sanvicente E, Yeoh GH et al (2012) Modelling of natural convection in vertical and tilted photovoltaic applications. Energy Build 55:810–822. Doi:10.1016/j.enbuild.2012.10.014

Lee B, Liu JZ, Sun B et al (2008) Thermally conductive and electrically insulating EVA composite encapsulant for solar photovoltaic (PV) cell. eXPRESS Polym Lett 2:357–363. Doi:10.3144/expresspolymlett.2008.42

Macq A, Mercier des Rochettes L, Martin-Carron L, et al (2016) A use of artificial intelligence for improving PV array performance (empirical approach). In: Proceedings of the 32nd European photovoltaic solar energy conference and exhibition

Maghanga CM, Niklasson GA, Granqvist CG, Mwamburi M (2011) Spectrally selective reflector surfaces for heat reduction in concentrator solar cells: modeling and applications of TiO_2:Nb-based thin films. Appl Opt 50:3296–302. Doi:10.1364/AO.50.003296

Makrides G, Zinsser B, Phinikarides A et al (2012) Temperature and thermal annealing effects on different photovoltaic technologies. Renew Energy 43:407–417. Doi:10.1016/j.renene.2011.11.046

Micheli L, Sarmah N, Luo X et al (2012) Infrared reflecting coverglasses for multijunction cells in a terrestrial high-concentrating photovoltaic system. In: Proceedings of the 27th European photovoltaic solar energy conference and exhibition. pp 266–270

Moser D, Pichler M, Nikolaeva-Dimitrova M (2013) Filtering procedures for reliable outdoor temperature coefficients in different photovoltaic technologies. J Sol Energy Eng 136:1–10. Doi:10.1115/1.4024931

Nardone M, Green B (2015) COMSOL-PV : a unified platform for numerical simulation of solar cells an modules. In: COMSOL conference proceedings

Naumann V, Lausch D, Hähnel A et al (2014) Explanation of potential-induced degradation of the shunting type by Na decoration of stacking faults in Si solar cells. Sol Energy Mater Sol Cells 120:383–389. Doi:10.1016/j.solmat.2013.06.015

Notton G, Cristofari C, Mattei M, Poggi P (2005) Modelling of a double-glass photovoltaic module using finite differences. Appl Therm Eng 25:2854–2877. Doi:10.1016/j.applthermaleng.2005.02.008

Odeh S, Behnia M (2009) Improving photovoltaic module efficiency using water cooling. Heat Transf Eng 30:499–505. Doi:10.1080/01457630802529214

Palyvos JA (2008) A survey of wind convection coefficient correlations for building envelope energy systems' modeling. Appl Therm Eng 28:801–808. Doi:10.1016/j.applthermaleng.2007.12.005

Pauling L (1988) General chemistry. Dover, Mineola, New York

Peike C, Hoffmann S, Hülsmann P et al (2013) Origin of damp-heat induced cell degradation. Sol Energy Mater Sol Cells 116:49–54. Doi:10.1016/j.solmat.2013.03.022

Prince MB (1955) Silicon solar energy converters. J Appl Phys 26:534–540. Doi:10.1063/1.1722034

Raman AP, Anoma MA, Zhu L et al (2014) Passive radiative cooling below ambient air temperature under direct sunlight. Nature 515:540–544. Doi:10.1038/nature13883

Rosell JI, Ibanez M (2006) Modelling power output in photovoltaic modules for outdoor operating conditions. Energy Convers Manag 47:2424–2430. Doi:10.1016/j.enconman.2005.11.004

Ross R, Gonzalez C (1980) Reference conditions for reporting terrestrial photovoltaic performance. In: AS/ISES annual meeting. pp 1091–1097

Royne A, Dey CJ, Mills DR (2005) Cooling of photovoltaic cells under concentrated illumination: a critical review. Sol Energy Mater Sol Cells 86:451–483

Rubin M (1985) Optical properties of soda lime silica glasses. Sol Energy Mater 12:275–288. Doi:10.1016/0165-1633(85)90052-8

Rus-Casas C, Aguilar JD, Rodrigo P et al (2014) Classification of methods for annual energy harvesting calculations of photovoltaic generators. Energy Convers Manag 78:527–536. Doi:10.1016/j.enconman.2013.11.006

Safi TST, Munday JJN (2015) Improving photovoltaic performance through radiative cooling in both terrestrial and extraterrestrial environments. Opt Express 23:A1120. Doi:10.1364/OE.23.0A1120

Schwingshackl C, Petitta M, Wagner JE et al (2013) Wind effect on PV module temperature: analysis of different techniques for an accurate estimation. Energy Procedia 40:77–86. Doi:10.1016/j.egypro.2013.08.010

Shademan M, Hangan H (2009) Wind loading on solar panels at different inclination angles. In: Proceedings of the 11th Americas conference on wind engineering

Silva JP, Nofuentes G, Munoz JV (2010) Spectral reflectance patterns of photovoltaic modules and their thermal effects. J Sol Energy Eng Asme 132:13. Doi:10.1115/1.4002246

Silvaco (2016) SILVACO. http://www.silvaco.com/products/tcad/device_simulation/atlas/atlas.html. Accessed 22 Aug 2016

Silverman TJ, Jahn U, Friesen G et al. (2014) Characterisation of performance of thin-film photovoltaic technologies (from Report IEA-PVPS T13-02:2014)

Skoplaki E, Palyvos JA (2009a) Operating temperature of photovoltaic modules: a survey of pertinent correlations. Renew Energy 34:23–29. Doi:10.1016/j.renene.2008.04.009

Skoplaki E, Palyvos JA (2009b) On the temperature dependence of photovoltaic module electrical performance: a review of efficiency/power correlations. Sol Energy 83:614–624. Doi:10.1016/j.solener.2008.10.008

Staebler DL, Wronski CR (1977) Reversible conductivity changes in discharge-produced amorphous Si. Appl Phys Lett 31:292–294. Doi:10.1063/1.89674

Strevel N, Trippel L, Gloeckler M, Solar F (2012) First solar : greater energy yields in high-temperature conditions performance characterization and superior energy yield of first solar PV power plants in high-temperature conditions. Photovoltaics Int 1–6

Tina GM, Rosa-Clot M, Rosa-Clot P, Scandura PF (2012) Optical and thermal behavior of submerged photovoltaic solar panel: SP2. Energy 39:17–26. Doi:10.1016/j.energy.2011.08.053

Trapani K, Redon Santafé M (2014) A review of floating photovoltaic installations: 2007–2013. Prog Photovoltaics Res Appl 23:524–532

Virtuani A, Pavanello D, Friesen G (2010) Overview of temperature coefficients of different thin film photovoltaic technologies. In: Proceedings of the 25th European photovoltaic solar energy conference

Weiss L, Amara M, Ménézo C (2016) Impact of radiative-heat transfer on photovoltaic module temperature. Prog Photovoltaics Res Appl 24:12–27

Witmer L (2010) Quantification of the passive cooling of photovoltaics using a green roof. PhD thesis

Wysocki JJ, Rappaport P (1960) Effect of temperature on photovoltaic solar energy conversion. J Appl Phys 31:571–578

Zhao J, Wang A, Campbell P, Green MA (1999) 22.7 % efficient silicon photovoltaic modules with textured front surface. IEEE Trans Electron Devices 46:1495–1497. Doi:10.1109/16.772498

Zhu L, Raman A, Wang KX et al (2014) Radiative cooling of solar cells. Optica 1:24–26. Doi:10.1364/OPTICA.1.000032

Chapter 2
Temperature Coefficients of Photovoltaic Devices

Abstract This chapter introduces the concept of temperature coefficient which enables to quantify the temperature sensitivity of the performances of photovoltaic devices. The temperature sensitivity of a photovoltaic converter originates from the temperature dependence of the fundamental conversion losses and of the bandgap of the absorber. The detailed balance principle is used to derive the fundamental losses which set the maximum of photovoltaic conversion efficiency in the radiative limit. It is highlighted that the unusual temperature dependence of the bandgap of perovskite compounds will ultimately result in peculiar temperature sensitivities. Following the discussion on the fundamental losses, the additional losses limiting the efficiency of present commercial cells are considered. The different loss mechanisms that drive the temperature coefficients of important cell parameters (open-circuit voltage V_{oc}, short-circuit current density J_{sc}, fill factor FF) are identified. The analysis shows how each of these temperature coefficients can provide some insight into device physics. The temperature sensitivity of open-circuit voltage is connected to the balance between generation and recombination of carriers and its variation with temperature. The temperature sensitivity of short-circuit current is driven by the bandgap temperature dependence and the incident spectrum on one hand and on the temperature dependence of the collection efficiency of the device on the other hand. As for the fill factor temperature sensitivity, it is ideally closely related to the open-circuit voltage temperature sensitivity but it also depends for certain devices on technological issues linked to carrier transport. The chapter ends with an overview of the possible approaches to tune the temperature coefficients of photovoltaic devices.

2.1 Definition

The concept of temperature coefficient (TC) is useful to quantify the temperature sensitivities of the performances of photovoltaic (PV) devices. In order to compare different technologies, TCs are defined normalized at 25 °C (298.15 K) (Emery et al. 1996):

O. Dupré et al., *Thermal Behavior of Photovoltaic Devices*,
DOI 10.1007/978-3-319-49457-9_2

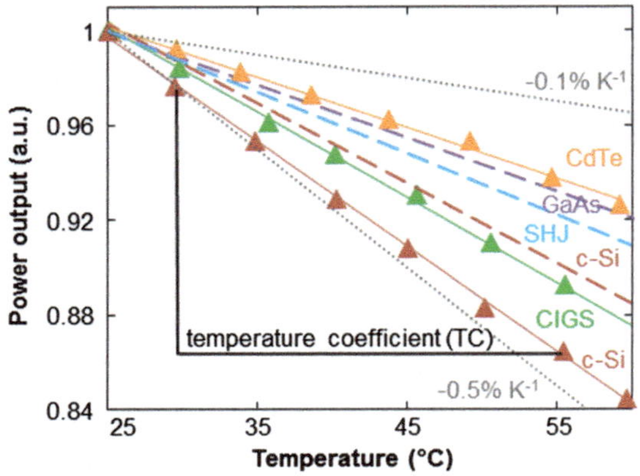

Fig. 2.1 Relative power output as a function of temperature of solar cells made of different semiconductors. Data from Siefer et al. (2005), Virtuani et al. (2010), Seif et al. (2014), Dupré et al. (2015a)

$$\beta_G(T_c) = \frac{10^6}{G(298.15\,\mathrm{K})} \frac{G(T_c) - G(298.15\,\mathrm{K})}{T_c - 298.15} \tag{2.1}$$

where G is the parameter of interest and T_c is the cell temperature. Temperature coefficients are usually expressed in ppm K^{-1} or in % K^{-1}. If variations of G are linear with temperature, β_G is well described by a single value. Conveniently, this is the case for certain important PV parameters (such as the maximum output power P_{MPP}, the open-circuit voltage V_{oc}, the short-circuit current density J_{sc} and the fill factor FF) of most types of PV cells (except a-Si cells, see Sects. 1.1.2 and 4.4, and certain other technologies such as perovskite cells, see Sect. 4.5). This is illustrated in Fig. 2.1 with the linear fits of the power outputs as a function of cell temperature for solar cells (SCs) made of different semiconductors. It is worth noticing that the efficiencies of all the PV devices shown in Fig. 2.1 decrease with increasing temperature. Also, the relative decrease rates (i.e. the temperature coefficients) are different among the semiconductors. There are also variations between devices made of the same semiconductor (e.g. c-Si in Fig. 2.1).

This chapter details the fundamental reasons for the negative temperature dependence of the photovoltaic conversion process and explains the important role of the bandgap and its temperature dependence in the thermal behavior of PV devices (Sect. 2.2). Then, Sect. 2.3 highlights the technological factors that drive the temperature coefficients of actual photovoltaic devices. Finally, Sect. 2.4 reviews the opportunities for tuning the temperature sensitivities of solar cells.

2.2 Fundamental Conversion Losses and Temperature Coefficients of Solar Cells

2.2.1 The Detailed Balance Principle and the Thermodynamic Argument

Fundamentally, solar cells are energy converters that take thermal energy from the Sun and turn it into electrical energy. This means that a solar cell, like any heat engine, is ultimately limited by the Carnot efficiency (which is ~95% since the Sun's surface temperature is ~5800 K) (Landsberg and Badescu 2000; Green 2001; Boriskina et al. 2016). However, even ideal PV devices differ from Carnot engines because the energy exchanged is radiative and because the energy emitted by the devices is considered a loss in photovoltaic conversion since the hot reservoir is the Sun (Fig. 2.2a). Moreover, typical PV cells absorb solar photons from a small solid angle while they emit in a much broader solid angle (Fig. 2.2b). Additionally, for standard single-junction cells, there is an important loss due to the spectral mismatch between the incident radiation and the absorption in the cell that generates electrical carriers (i.e. single bandgap devices absorb inefficiently the

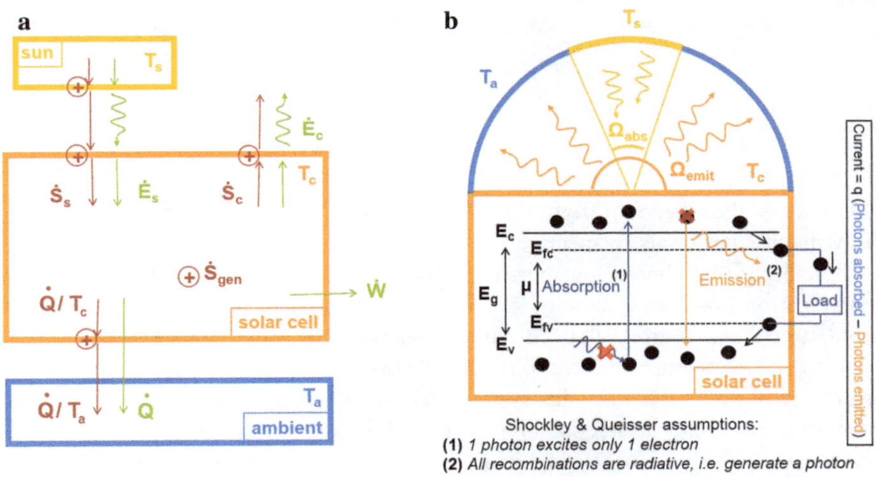

Fig. 2.2 **a** Flows of energy (in *green*) and entropy (in *red*) in and out of a solar cell. A solar cell absorbs radiative energy from the Sun with the associated entropy (see (Green 2001) for the corresponding expressions). It also emits some radiative energy with the associated entropy and evacuates some heat with the associated entropy towards the ambient. The symbol + represents the generation of entropy. The equilibrium temperature of the cell is driven by the balance between the entropy flows in and out of the cell. The power with no associated entropy flow can be converted into electrical power ($\dot{W}$); **b** Detailed balance principle illustrated with the photons fluxes in and out of a schematic solar cell. This schematic shows the direct relation that exists in the radiative limit between the populations of excited carriers and photons. It also specifies the assumptions made by Shockley and Queisser to derive their well-known efficiency limit (Shockley and Queisser 1961)

broadband incident solar spectrum). The present section illustrates how these different losses impact the conversion efficiency of PV systems.

The first assessment of the fundamental limit of photovoltaic conversion, established by Shockley and Queisser in 1961, uses the so-called detailed balance principle (Shockley and Queisser 1961). This limit, extended by several authors (Henry 1980; Tiedje et al. 1984; Araújo et al. 1994), can also be derived from a thermodynamic argument (Ross 1967; Markvart 2008). In their well-known article, Shockley and Queisser consider that all the recombinations in the solar cell are radiative, i.e. each recombination event creates a photon, and assume that each photon excites only one electron so that the population of photons and excited electrons are directly related (Fig. 2.2b). In this case, the electrical current density (J) that can be extracted from a solar cell is proportional to the difference between the rates of photon absorption (n_{abs}) and emission (n_{emit}):

$$J = q\left(n_{abs} - n_{emit}\right). \tag{2.2}$$

where q is the elementary charge. The source of incident photons, the Sun in the case of a solar cell, can be modeled by a blackbody at temperature T_s = 5800 K seen by the cell from a solid angle $\Omega_{abs.}$[1] It is assumed that the cell absorbs, and thus emits according to Kirchhoff's law, every photon that has at least the bandgap energy (E_g) and none that has less energy. The rate of photon absorption can be formulated by integrating Planck's equation for blackbody emission:

$$n_{abs}(E_g, T_s, \Omega_{abs}) = \frac{2\,\Omega_{abs}}{c^2 h^3} \int\limits_{Eg}^{\infty} \frac{E^2}{\exp\left(\frac{E}{kT_s}\right) - 1} dE \tag{2.3}$$

where c is the speed of light in vacuum, and h and k are the Planck's and Boltzmann's constants, respectively.

Assuming that charge transport is perfect, the free enthalpy of the photogenerated electron-hole pairs, namely the chemical potential μ of the electron-hole system (Ruppel and Würfel 1980), is equal to qV where V is the voltage across the cell terminals. The chemical potential of the electron-hole system indicates a deviation from equilibrium. Because the system strives to come back to equilibrium, the recombination rate, which is directly linked to the emission rate in this case, increases together with the chemical potential. Considering a cell temperature T_c

[1]More precisely, here and everywhere in the present manuscript, the denomination solid angle Ω corresponds to a projection of the usual definition of the solid angle: $\int \cos\theta \, d\Omega = \int_0^{2\pi} \int_0^{\alpha} \cos\theta \, \sin\theta \, d\theta d\varphi$ (where θ and φ are the polar and azimuthal angles, respectively, and α is the angle of the cone of light) because an isotropic intensity is assumed. Note that, in normal incidence, this value is approximately equal to the solid angle if it is sufficiently small (e.g. the solid angle of the Sun seen from Earth is $6.87 \times 10^{-5} \approx \Omega_{abs} = \int \cos\theta \, d\Omega$).

and an emission solid angle Ω_{emit}[2], the photon emission rate from the cell is given by the generalized Planck's equation (Würfel 1982):

$$n_{emit}(E_g, V, T_c, \Omega_{emit}) = \frac{2\,\Omega_{emit}}{c^2\,h^3} \int_{Eg}^{\infty} \frac{E^2}{\exp\left(\frac{E-q\,V}{kT_c}\right) - 1} dE. \tag{2.4}$$

Equations (2.2) and (2.4) show that emission increases with the voltage V that builds in the cell and thus reduces the current that can be collected. As a consequence, there exists an optimum for the photovoltaic conversion of the radiative power incident on the cell (P_{inc}) noted η_{MPP} (MPP stands for maximum power point):

$$\eta_{MPP}(E_g, V, T_s, T_c, \Omega_{abs}, \Omega_{emit}) = \max(JV)/P_{inc}. \tag{2.5}$$

The maximum efficiency corresponds to a trade-off between current and voltage and depends on the difference between the rates of generation and recombination (in the radiative limit, the recombination rate equals the emission rate). In a given configuration (i.e. T_s, T_c, Ω_{abs}, Ω_{emit} are prescribed), the cell efficiency is only a function of E_g and V. Thus, its maximum, η_{MPP}, can be found by solving the following equations:

$$\left(\frac{\partial \eta}{\partial E_g}\right)_V = 0 \tag{2.6}$$

$$\left(\frac{\partial \eta}{\partial V}\right)_{E_g} = 0. \tag{2.7}$$

An interesting analytical solution of (2.6) can be derived by using Boltzmann's approximation[3] in (2.3) and (2.4) and calculating the integrals by parts (Hirst and Ekins-Daukes 2011):

$$q\,V_{opt} = E_g\left(1 - \frac{T_c}{T_s}\right) - k\,T_c\,\ln\left(\frac{\Omega_{emit}}{\Omega_{abs}}\right) \approx q\,V_{MPP}. \tag{2.8}$$

This analytical solution (V_{opt}) gives exactly the maximum efficiency only for the optimal E_g (i.e. when (2.7) is satisfied too) but it provides an excellent approximation of the voltage at the maximum power point (V_{MPP}) in the range of bandgaps of

---

[2]Note that this projected solid angle is equal to ½ of the solid angle in the case of an hemisphere (this is why usually $\Omega_{emit} = \int \cos\theta\, d\Omega = ½(2\pi) = \pi$).

[3]Photons enter the class of particles that follows Bose-Einstein statistics (these particles are called bosons). Boltzmann's approximation consists in considering the simpler Boltzmann statistics [i.e. neglecting the −1 term in the denominator of the integrand in (2.3)]. This approximation is really accurate in the range of bandgaps considered here: it is better than one part in 10^8 for $E_g > 0.5$ eV.

materials used in solar cells (Hirst and Ekins-Daukes 2011). In this equation, derived from the detailed balance principle, classical thermodynamic terms appear. The first term on the right hand side, $E_g\,(1 - T_c/T_s)$, contains the Carnot efficiency that expresses the necessity of evacuating the entropy associated with the absorbed energy flow. The second term on the right hand side, $k\,T_c\,\ln(\Omega_{emit}/\Omega_{abs})$, corresponds to the voltage loss related to the entropy generation caused by the solid angle mismatch between absorption and emission. This loss is theoretically avoidable by eliminating this solid angle mismatch. Interestingly, this can be achieved either by concentrating the incident light, i.e. increasing the solid angle of solar absorption, or by restricting the angle of emission from the device (Martí and Araújo 1996; Kosten et al. 2013, 2015). It is worth noting that these losses are function of the cell bandgap, temperature and solid angles of emission and absorption.

The current density at the maximum power point (J_{MPP}) is given by:

$$J_{MPP} = q\,(n_{abs} - n_{emit}(V_{MPP})). \tag{2.9}$$

Table 2.1 shows the analytical expressions of the energy losses related to the effects mentioned previously. The first two losses are related to the spectral mismatch between the broadband incident radiation and the restricted cell absorption: some photons have more energy than the bandgap and this "extra" energy is quickly lost by the excited electrons to the lattice atoms in a process called thermalization; some photons have less energy than the bandgap and are not able to excite any electron (sub-bandgap loss). The last three losses are driven by the balance between absorption and emission rates and thus are called "balance losses". The emission loss corresponds to the energy of the photons emitted by the PV device. The Carnot and angle mismatch losses are voltage losses associated with the emission process.

Figure 2.3a shows the fundamental losses, for single-junction solar cells under one Sun illumination, calculated as a function of bandgap energy from the analytical approximate expressions together with the exact numerical solution of the Shockley-Queisser (SQ) limit. This representation, first introduced in Hirst and Ekins-Daukes (2011), makes it straightforward to visualize the potential of different advanced PV concepts. Indeed, Fig. 2.3a highlights that the largest losses are those linked to spectral mismatch which explains the number of concepts aiming at

Table 2.1 Fundamental conversion losses in a single-junction solar cell

Spectral mismatch	Thermalization	$\frac{2\,\Omega_{abs}}{c^2\,h^3}\int_{Eg}^{\infty}\frac{E^2}{\exp\left(\frac{E}{kT_s}\right)-1}(E-E_g)\,dE$	(2.10)
	Sub-bandgap	$\frac{2\,\Omega_{abs}}{c^2\,h^3}\int_{0}^{E_g}\frac{E^2}{\exp\left(\frac{E}{kT_s}\right)-1}E\,dE$	(2.11)
Absorption-emission balance	Emission	$\frac{E_g}{q}\frac{2\,\Omega_{emit}}{c^2\,h^3}\int_{E_g}^{\infty}\frac{E^2}{\exp\left(\frac{E-qV_{MPP}}{kT_c}\right)-1}\,dE$	(2.12)
	Angle mismatch	$\frac{k\,T_c}{q}\ln\left(\frac{\Omega_{emit}}{\Omega_{abs}}\right)J_{MPP}$	(2.13)
	Carnot	$\frac{E_g}{q}\left(\frac{T_c}{T_s}\right)J_{MPP}$	(2.14)

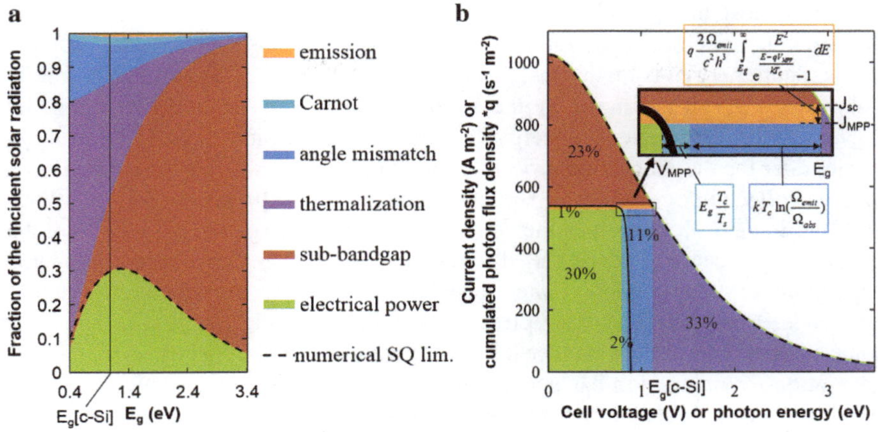

Fig. 2.3 **a** Distribution of the incident solar radiation at the optimal working point for PV conversion at 300 K under one sun illumination as a function of bandgap energy; the dashed line corresponds to the numerical solution of the Shockley-Queisser limit; **b** fundamental losses in PV conversion at the maximum power point of a c-Si PV cell at 300 K under one Sun illumination depicted together with its current-voltage characteristic

reducing these losses: hot-carrier cells (Ross and Nozik 1982; Green 2001), down conversion/multiple exciton generation (Nozik 2008; Beard 2011), intermediate band cells (Luque and Martí 1997; Brown and Green 2002; Luque et al. 2012), up conversion (Trupke et al. 2002), multi-junctions (Henry 1980), spectral splitting (Green et al. 2015), etc. Other strategies, such as concentrating sunlight (Press 1976) or limiting the cell emission angle (Kosten et al. 2013), improve the absorption-emission balance and thus reduce the angle mismatch loss (and emission as well in the "restricted angular emission" approach).

Figure 2.3b represents these fundamental loss mechanisms at the maximum power point together with the ideal current density-voltage (J-V) characteristic of a crystalline silicon (c-Si) solar cell at 300 K ($E_g \approx 1.12$ eV). The dashed line corresponds to the cumulated photon flux density plotted as a function of photon energy. The area below this line corresponds to the total power density of the incident photons. In the following, for the sake of simplicity, the term "energy" will be used instead of "power density" (energy simply corresponds to the spatial and temporal integral of power density). The different colored areas in Fig. 2.3b correspond to the different energy conversion losses in an ideal c-Si cell except for the pale green rectangle that corresponds to the electrical output energy. The dual title on the y-axis of Fig. 2.3b illustrates the relation between current and photon flux density. Indeed, each absorbed photon generates one carrier that can participate, if it does not recombine on its way, to the current that goes through the load. In short circuit, the current density, J_{sc}, is equal to the cumulated photon flux density at the bandgap because the radiative emission (and thus the recombination rate) is negligible in short-circuit condition. When the voltage of the cell increases, so does the radiative emission rate and the current drops.

The energy lost through radiative emission at the maximum power point corresponds to the number of emitted photons times their energy which is approximately equal to the bandgap energy. As for the dual title on the x-axis of Fig. 2.3b, it illustrates that the voltage of the cell is limited by the bandgap energy for standard single-junction devices.[4] Indeed, the potential energy of an excited carrier cannot exceed the energy it carries after the thermalization process because this energy loss happens much faster that the collection of excited carriers in common PV cells [in about 10^{-12} s (Würfel 2009)]. Figure 2.3b shows that the voltage at the maximum power point is lower that the bandgap because of the so-called Carnot and angle mismatch losses. These terms correspond to the minimum voltage reduction associated with radiative recombination. Indeed, any recombination process has a dual impact: a current loss because some excited charges do not make it to the external circuit; a voltage loss because the generation-recombination balance is diminished thus reducing the voltage that can build in the cell. An analogy between a PV cell and a leaky bucket feeding a water wheel is proposed in Appendix 1 to provide an intuitive understanding of the relation between the cell voltage and the generation-recombination balance. The Carnot loss corresponds to the voltage loss associated with the minimum radiative recombination rate, which happens in the radiative limit when there is no angle mismatch between radiative absorption and emission. When there is an angle mismatch, there is more emission[5] so there has to be more radiative recombination and thus a larger voltage loss, hence the angle mismatch loss term. It is important to remember that the expressions given in Table 2.1 describe only the maximum power point of the cell operation. However, it can be deduced from Fig. 2.3b how the incident energy is distributed in other $J(V)$ configurations.

The p-n junction diagram depicted in Fig. 2.4 can also help understanding the relation between the J-V curve and the different losses indicated in Fig. 2.3. In open-circuit configuration, the voltage across the cell is such that the recombination rate equals the generation rate: all the absorbed power that is not thermalized ($\sim J_{sc}\ E_g/q$) is dissipated via radiative recombinations which ultimately result in the emission of photons. In short-circuit configuration, there is no voltage across the cell so almost none of the photogenerated carriers recombine. All the absorbed power that is not thermalized is dissipated by the carrier flow through the potential drop at the junction ($\sim E_g/q$ in short circuit). At the maximum power point (MPP), there are some recombinations (current loss = emission loss term) and the voltage loss at the junction depends on the recombination rate which is driven by the cell temperature and the angle mismatch.

Figure 2.4 depicts how the energy of the incident photons is converted within an ideal p-n junction solar cell. The voltage losses can be observed on the vertical axis while the current losses are depicted by photon losses along the horizontal axis. The sub-bandgap loss represents the photons with $E < E_g$. The dashed lines show the quasi-Fermi levels of the electrons and holes. The distances between the

[4]Note that this limit can be overcome through advanced concepts such as hot-carrier cells.

[5]Kirchhoff's law of radiation dictates that $\Omega_{emit} \geq \Omega_{abs}$.

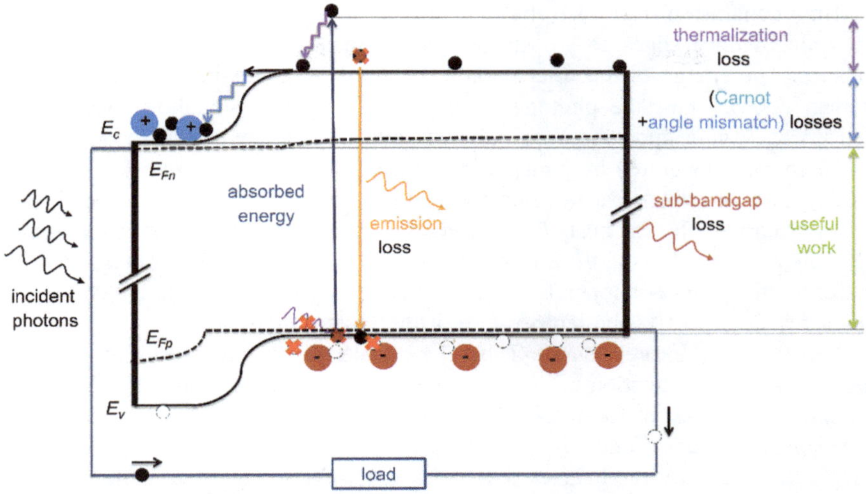

Fig. 2.4 Fundamental loss mechanisms illustrated on a p-n junction diagram. E_c and E_v are the energies of the bottom edge of the conduction band and the top edge of the valence band, respectively. E_{Fn} and E_{Fp} are the quasi-Fermi levels of electrons and holes, respectively. Figure adapted from Dupré et al. (2015b)

quasi-Fermi levels and the conduction/valence bands give an indication on the concentrations of carriers in these bands (Green 1982). This representation shows that the Carnot and angle mismatch losses occur at the p-n junction or more generally where the charges are separated (see also Fig. 2.11). This voltage loss can sometimes be split into different regions, e.g. in p-i-n junction cells. This voltage drop is necessary to efficiently collect the photogenerated charges before too many of them recombine. Physically, a fraction of their potential energy is converted into kinetic energy during the acceleration they undergo in the junction electric field. Then, they quickly relax to the potential energy of the conduction band of the other side through collisions with the lattice atoms. This heat generation process can be identified as Peltier heating (Baldasaro et al. 2001), as discussed in Chap. 3.

The fundamental losses of PV conversion being established, their dependence on temperature is the next focus. Equation (2.4) shows that radiative emission increases with cell temperature. This is the consequence of an augmentation of the recombination rate with temperature due to the increase of the equilibrium carrier concentration. This leads to a negative temperature sensitivity of the absorption-emission rate balance that is the origin of the thermal behavior of PV cells and explains why all[6] PV devices become less efficient as they heat up.

[6]Note that positive temperature coefficients have been reported for amorphous silicon and organic solar cells. However, the argument developed above holds for any kind of PV conversion device. The temperature coefficients of the aforementioned cells are expected to decrease and become negative as these technologies improve towards the radiative limit.

This conclusion can also be derived from a thermodynamical argument. Thermodynamics states that a stable system needs to have a zero net entropy balance. The energy of the solar photons comes with a certain amount of entropy (Green 2006). Thus, the photogenerated carriers need to somehow dissipate the entropy they gain upon the absorption of solar photons and possibly some additional entropy generated by parasitic processes. They do so by giving away some energy as heat, which carries entropy, to the lattice atoms. Because the flow of entropy carried by a heat flow decreases when the temperature increases, $\delta S = \delta Q/T$, equilibrating the entropy balance requires a larger loss of energy to the lattice atoms at higher temperature. This explains why the maximum PV conversion efficiency decreases as the temperature of the system increases.

The analytical expressions in Table 2.1 enable to calculate the different fundamental losses as a function of device temperature. These are plotted in Fig. 2.5 for the bandgap of c-Si at room temperature (~ 1.12 eV). Figure 2.5a corresponds to one Sun illumination and Fig. 2.5b to a concentration factor of 10,000. In this simplistic case, only the balance losses, i.e. emission, Carnot and angle mismatch losses, are sensitive to the cell temperature (see Table 2.1). Figure 2.5 shows that their variations with temperature are approximatively linear on the usual temperature range of solar cell operation. This explains the generally observed linear behavior of the output power of solar cells. Also, by comparing Fig. 2.5a and b, the benefit of concentrating sunlight in terms of temperature coefficient can be observed.

The analytical expressions in Table 2.1 show that the balance losses depend not only on the cell temperature but also on the semiconductor bandgap. This means that the temperature coefficient of PV conversion is fundamentally dependent upon the bandgap of the material used to absorb solar radiation. Figure 2.6 shows temperature coefficients in the radiative limit calculated as a function of cell bandgap with the analytical expressions and the full numerical calculations using the detailed balance approach. It is noteworthy that the analytical solution provides a good

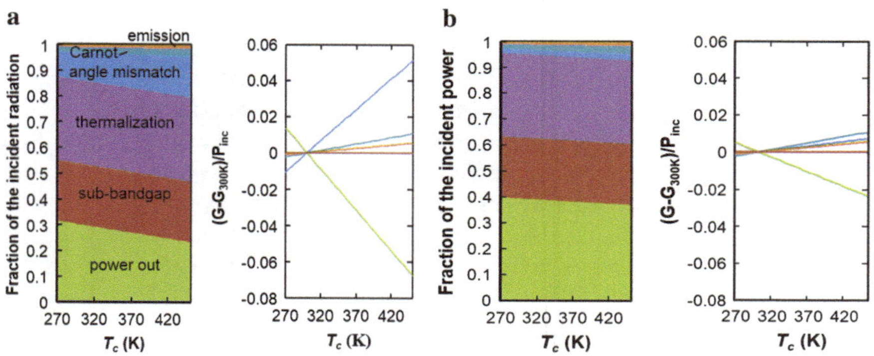

Fig. 2.5 Temperature dependence of the fundamental losses for a bandgap of 1.12 eV ($\sim E_g$(c-Si) at 298 K), **a** under one Sun illumination; **b** under 10,000 Suns. Figure adapted from Dupré et al. (2015b)

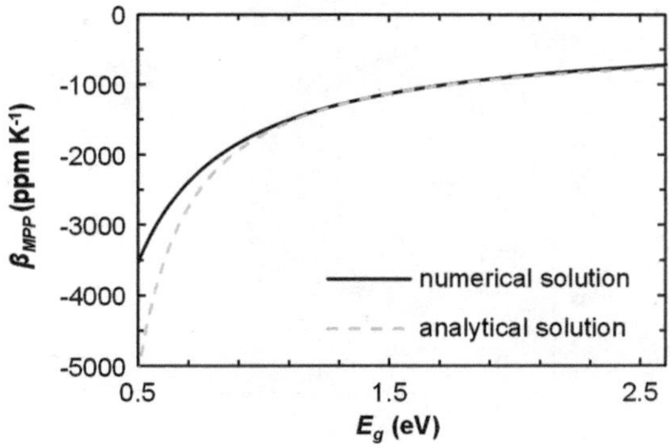

Fig. 2.6 Fundamental temperature coefficient of the maximum efficiency of single-junction solar cells as a function of bandgap

estimate of temperature coefficients as long as E_g is not too small. Also, Fig. 2.6 highlights that the absolute value of the temperature coefficient is larger for smaller bandgaps. This is due to the fact that balance losses, which increase with temperature because they originate from photon emission, are larger for smaller bandgaps (see Fig. 2.3a).

2.2.2 *Influence of Bandgap Temperature Dependence and Incident Spectrum*

All the fundamental losses in PV conversion are function of the bandgap of the material used to absorb solar radiation (see Table 2.1). The temperature coefficients plotted in Fig. 2.6 assume that bandgaps do not depend on temperature. In practice, it is well known that the bandgap of semiconductors changes substantially with temperature. This variation is due to modifications of the band energies caused by electron-phonon interactions and by thermal expansion of the lattice (Dey et al. 2013). Most semiconductor bandgaps decrease almost linearly in the temperature range of operation of solar cells (Yu et al. 2011). However, there is no general relation between bandgap and temperature dependence of bandgap as shown in Fig. 2.7.[7] Also, there are some exceptions where bandgaps actually increase with temperature. Of particular interest for photovoltaics are the perovskite[8] semiconductor compounds such as $CH_3NH_3PbI_3$ (Jiang et al. 2016), $CH_3NH_3PbI_{3-x}Cl_x$

[7]Note that there are several discrepancies for different semiconductors in the values of $E_g(T)$ reported in the literature (Varshni 1967; Pässler 1999; Vurgaftman et al. 2001).

[8]Perovskites are materials described by the formula ABX_3 where A and B are cations of different sizes (A larger than B) and X is an anion. A new class of solar cells based on mixed organic-inorganic halide perovskites (such as $CH_3NH_3PbI_{3-x}Cl_x$) has recently emerged and undergoes an extraordinary rapid development (Green et al. 2014). The inorganic counterpart $CsSnI_3$ has only achieved very low efficiency so far for reasons that are not understood yet, but also has a suitable bandgap for PV applications (Xu et al. 2014).

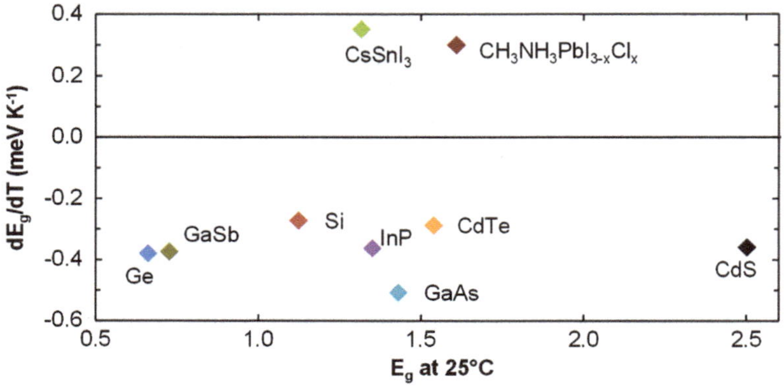

Fig. 2.7 Bandgap and slope of the linear approximation of bandgap temperature dependence around 300 K for various semiconductors. See Table 2.2 for the references

(Wu et al. 2014) or $CsSnI_3$ (Yu et al. 2011). A notable feature of perovskites is a reverse ordering of their band-edges states, i.e. their conduction band is constituted of p-like states and their valence band of s-like states (Even et al. 2012). This reverse band-edge ordering results in a bandgap temperature dependence opposite to that of tetrahedrally coordinated semiconductors (Ishihara 1994; D'Innocenzo et al. 2014; Green et al. 2014).

Since the maximum efficiency of photovoltaic conversion depends on bandgap, the temperature dependence of semiconductor bandgaps impacts directly the temperature coefficients. Also, since bandgaps vary with temperature, the spectrum of the incident radiation has an effect on the temperature coefficients. Indeed, the number of photons with energy larger than the bandgap changes with temperature as a function of bandgap variation and photon flux density. These effects are illustrated in Figs. 2.8 and 2.9. Figure 2.8 shows the Shockley-Queisser efficiency limit calculated with the AM1.5 spectrum (IEC 2008) at different temperatures with the maximum efficiencies for different semiconductors. It illustrates the different behaviors expected for the perovskites compounds whose bandgaps increase with temperature. In order to illustrate the impacts of the incident spectrum and the bandgap temperature dependences, the dashed lines in Fig. 2.9 show extrapolations of temperature coefficients as a function of bandgap assuming different values of dE_g/dT. Two extremes values of dE_g/dT (for this set of semiconductors) are used for illustration so the temperature coefficients of the materials considered are sandwiched between the dashed lines. The bumps visible on the dashed lines in Fig. 2.9 correspond to the bumps in the plots of maximum efficiency as a function of bandgap (Fig. 2.8) which originate from the irregularities in the incident AM1.5 spectrum caused by the absorption of certain molecules and Rayleigh scattering in the atmosphere (Fig. 2.10). Because of these irregularities, the impact of dE_g/dT on the maximum power temperature coefficient depends a lot on the value of the bandgap at 25 °C. Also, two semiconductors with similar bandgaps at room temperature but different dE_g/dT will ultimately, i.e. in the radiative limit, have different thermal behaviors. For example, a PV cell made of $CH_3NH_3PbI_3$ will have a worse temperature coefficient in the radiative limit than cells made of semiconductors with a similar bandgap but a positive dE_g/dT. However, the temperature coefficients of actual solar cells depend largely on their external radiative efficiencies (see Sect. 2.3) which are already relatively high for the best perovskite solar cells to date (Green et al. 2014).

Figures 2.8 and 2.9 show that the impact of dE_g/dT on temperature coefficient is function of $d\eta/dE_g$ and thus depends on E_g and the incident spectrum. This explains

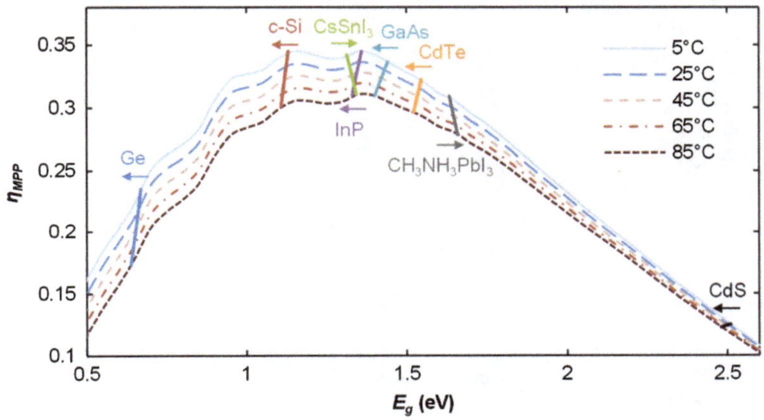

Fig. 2.8 Shockley-Queisser limit at different temperatures between 5 and 85 °C. The segments correspond to the maximum theoretical efficiencies for different semiconductors calculated using bandgap temperature dependences (references are listed in Table 2.2). Figure adapted from Dupré et al. (2015b)

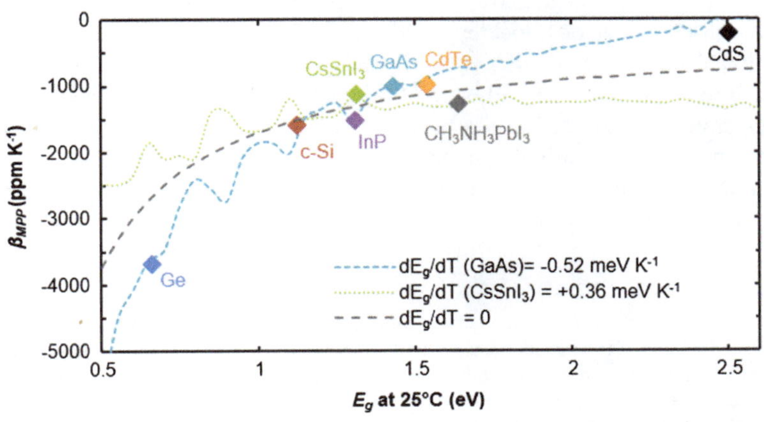

Fig. 2.9 Temperature coefficients of several semiconductors and extrapolations using different values of dE_g/dT. Figure adapted from Dupré et al. (2015b)

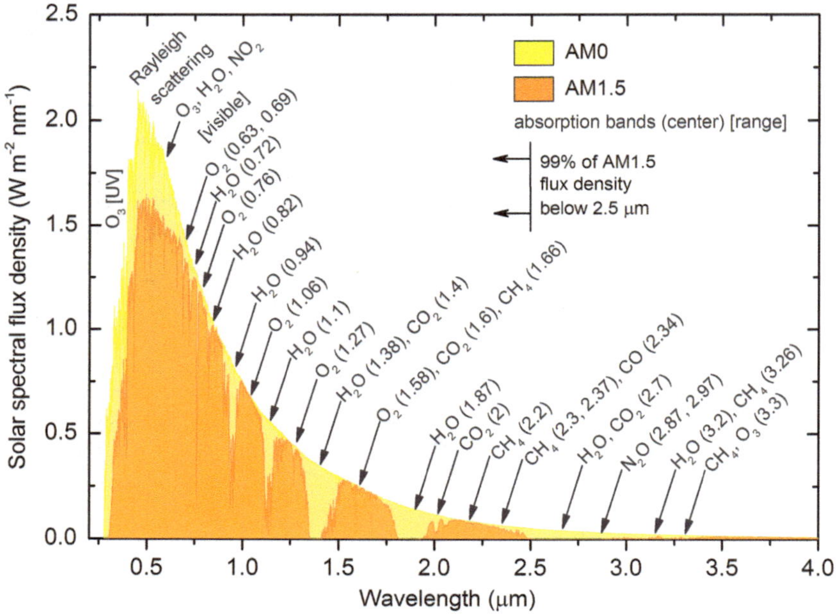

Fig. 2.10 Solar spectral flux density at the top of the atmosphere (AM0) and at sea level at a solar zenith angle of 48.19°S (AM1.5) from IEC (2008). The atmospheric absorption bands are taken from Liou (2002)

why dE_g/dT is of particular importance for semiconductors whose bandgaps are away from the optimum of the SQ limit (such as several sub-cells of multi-junction PV devices).

So far, the analysis has been derived in the radiative limit where only radiative recombinations are considered, i.e. neglecting non-radiative recombination processes such as Auger or Shockley-Read-Hall. While this situation corresponds to the ultimate efficiency of solar cells, it does not give the theoretical minima of temperature coefficients. Indeed, some of the additional losses may decrease when the temperature rises [e.g. series resistance in c-Si cells (Singh et al. 2008)]. This is the reason why unusually large or even positive[9] temperature coefficients are sometimes reported for newly-developed technologies (Green 2003). However, temperature coefficients are expected to converge towards these "fundamental"

[9]Several articles report positive temperature coefficients of the maximum power output for organic solar cells (Katz et al. 2001; Riedel et al. 2004; Ali et al. 2012). This is due to important losses at room temperature that are reduced as the temperature rises. According to Katz et al., the large TCs of cells made of conjugated polymer-fullerene composites originate from the temperature dependence of their carrier mobilities (Katz et al. 2001). Therefore the TCs of organic solar cells are expected to decrease as the technologies improve, which is the opposite of the trend of most inorganic type of solar cells.

values as technologies improve towards the radiative limit (this is illustrated for silicon cells in Fig. 2.14). To be exact, Auger recombinations[10] have to be taken into account when calculating the intrinsic limit for PV conversion of a semiconductor (Richter et al. 2013). For some semiconductors, this impacts significantly the temperature coefficients. For example, the intrinsic temperature coefficient for crystalline silicon solar cells is −2380 ppm K^{-1} while it is −1582 ppm K^{-1} in the radiative limit (Dupré 2015). This is to be compared to the values around −4500 ppm K^{-1} (0.45% °C^{-1}) that are common for the temperature coefficients of current commercial c-Si solar cells. The following section describes the loss mechanisms that limit the efficiency of actual PV devices and highlights those that impact their temperature coefficients.

2.3 Loss Mechanisms and Temperature Coefficients of Actual Solar Cells

Present commercial photovoltaic systems have efficiencies significantly lower than the Shockley-Queisser limit defined by the fundamental losses described previously. For example, commercial silicon solar modules on the market in 2016 have, on average, rated efficiencies of about 17% (Fraunhofer 2016) while the SQ limit for c-Si is 33.4%. Note that the record laboratory c-Si cell efficiency, which is 25.6% in July 2016 (Green et al. 2016), is getting close to the intrinsic limit efficiency of 29.43% for c-Si (Richter et al. 2013). Furthermore, the gap between the efficiency of commercial c-Si cells and the record laboratory cells is rapidly closing (Green, Sect. 2 in Boriskina et al. 2016). In this section, the additional losses limiting real device performances are introduced and their impacts on temperature coefficients are analyzed.

Figure 2.11 shows the band diagram of a realistic p-n junction solar cell operating at its maximum power point. The most important losses usually stem from the non-radiative recombinations (NRR). The different NRR processes (Shockley-Read-Hall,[11] Auger, surface[12]) are illustrated. Other losses include

[10]An Auger recombination corresponds to the crossing of an electron from the conduction band to the valence band via the transfer of some energy (at least E_g) to another electron of the conduction band. The electron of the conduction band that receives the energy quickly thermalizes back to the lower energy level of the conduction band (E_c). This process is illustrated on the p-n junction diagram in Fig. 2.11. It does not depend on defects but mainly on the local concentration of carriers, electrons and holes, in the conduction and valence bands.

[11]Shockley-Read-Hall (SRH) recombination is an extrinsic two step process by which an excited carrier recombines trough a defect state in the forbidden region (i.e. in the bandgap). Defects in the crystal lattice can have several origins such as the presence of impurities (unintentional or intentional, e.g. dopants). SRH recombination depends strongly on defect type and can have various dependences on temperature (Schenk 1992).

[12]Surfaces are regions where the lattice structure is bound to be filled with defects. Additionally, impurities tend to gather in these regions. Consequently, the bandgap is filled with defect states

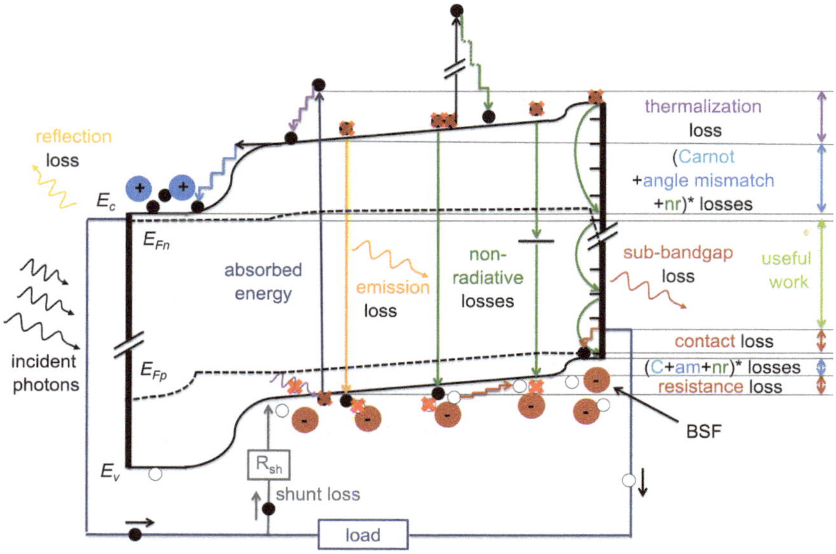

Fig. 2.11 Conversion loss mechanisms of a PV cell illustrated on a p-n junction diagram at the maximum power point. Figure adapted from Dupré et al. (2015b)

reflection at the front—illuminated side—of the cell, electrical shunts, imperfect contacts and finite mobilities of the carriers. Note that transmission losses, i.e. photons with $E > E_g$ that are not absorbed, are not depicted here but can also reduce the efficiency of thin PV cells with insufficient light trapping. The graph depicted in Fig. 2.11 is interesting in that it illustrates the different energy transformations in a solar cell. For example, it shows that the resistance loss corresponds to the kinetic energy lost by the electrical carriers to the semiconductor or metal atoms through collisions along their paths.

Similarly to Fig. 2.4, Fig. 2.11 shows that the mechanisms reducing the generation-recombination balance result in a voltage loss which occurs at the interfaces with the selective membranes that drive the photogenerated charges in opposite directions. In this example, the interfaces are the p-n junction and also the region of the back surface field (BSF) which can be seen as a p-p + junction. For other configurations, charge separation can be different (e.g. in p-i-n structures, this voltage loss is split more equally between the two separate junctions). Reflection, transmission and shunts, similarly to any current loss, also add to this voltage drop but their contribution is usually negligible so it is not depicted in Fig. 2.11 (and subsequently in Fig. 2.12). The product of the current and this potential drop

(Footnote 12 continued)

and recombinations rates are usually very high. A parameter called "surface recombination velocity" is used to quantify recombination rates at surfaces.

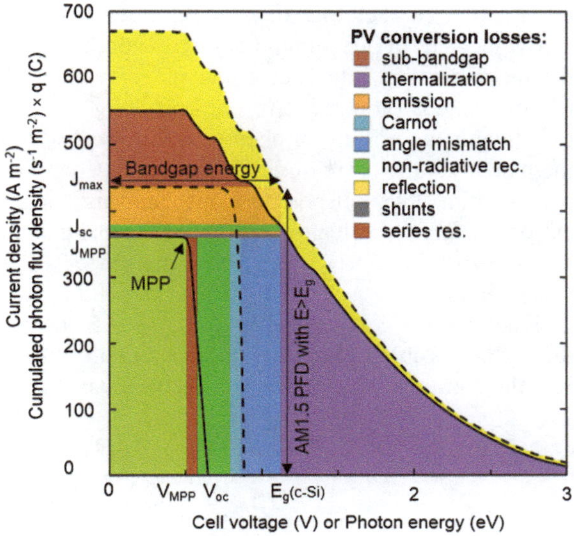

Fig. 2.12 Conversion loss mechanisms of a c-Si cell at MPP. Cumulated photon flux density of the AM1.5 spectrum (*dashed line*), of the transmitted fraction of this incident radiation perpendicularly through a planar c-Si surface (*solid line*). Ideal (*dashed line*) and actual (*solid line*) J-V characteristics of a c-Si solar cell with a planar surface. Figure adapted from Dupré et al. (2015b)

corresponds to a "collection cost": the energy required to extract the photogenerated electrical carriers from the cell before too many of them recombine.

In Fig. 2.12, the J-V curve of an illuminated c-Si cell is depicted together with the different losses at the maximum power point. Figure 2.12 shows together the cumulated photon flux density as a function of photon energy and the current density as a function of voltage (see the description of Fig. 2.3b). While Fig. 2.3b shows the cumulated photon flux density of a blackbody emission spectrum at 5800 K, Fig. 2.12 shows the cumulated photon flux density of the AM1.5 spectrum (dashed line). The full line corresponds to the cumulated photon flux density absorbed by a planar crystalline silicon cell.[13] The area between these two cumulated photon flux densities corresponds to the total energy[14] of the reflected photons. The rectangle in pale yellow illustrates the energy lost for the conversion process, i.e. the bandgap energy times the number of reflected photons with $E > E_g$. Note that the reflection loss in actual c-Si cells is much lower because of the use of anti-reflection coatings and texturization.

Figure 2.12 illustrates the different "fundamental" losses, presented in the description of Fig. 2.3b, together with the additional losses that occur in actual solar cells as described above and depicted in Fig. 2.11. As in Fig. 2.3b, by identifying the losses that limit current and voltage, it is possible to understand the shape of the J-V characteristic of a solar cell. In short-circuit configuration, there is an important current flow and no voltage across the cell: all the absorbed power is dissipated

---

[13]Calculated with reflectivity data from OPALv1.3 (Baker-Finch and McIntosh 2010).

[14]In this paragraph, for the sake of simplicity, "power densities" are once again replaced by "energies".

because the charges lose their potential energy to the phonons in voltage drops across the cell. In open-circuit configuration, the voltage across the cell is such that the recombination rate equals that of generation: all the power is dissipated via recombinations (non-radiative ones resulting in heat sources and a fraction of the radiative ones resulting in photon emission). The differences in heat generation in these configurations are discussed further in Chap. 3 (Sect. 3.4.1). At the maximum power point, the transport loss mechanisms come into play because there is a current flow and a voltage across the cell. Note that the dependence of the shunt and series resistances on voltage drive the slopes of the two segments of the J-V characteristic.

Since the PV cell parameters (V_{oc}, J_{sc}, FF) usually vary linearly with temperature, it is possible to separate the temperature sensitivity of a device performance into the sum of their temperature coefficients:

$$P_{MPP}(T_c) = V_{oc}(T_c)\ J_{sc}(T_c)\ FF(T_c) \tag{2.15}$$

$$\beta_{P_{MPP}} = \beta_{V_{oc}} + \beta_{J_{sc}} + \beta_{FF} \tag{2.16}$$

where P_{MPP} is the power density at the maximum power point. This is particularly interesting because these different TCs depend on different loss mechanisms. Thus, these temperature dependences provide some insight into internal device physics.

2.3.1 Open-Circuit Voltage Temperature Sensitivity

The temperature sensitivity of open-circuit voltage is of particular importance because it accounts for 80–90% of the temperature coefficient of efficiency for reasonably good solar cells (Green 2003).

The open-circuit configuration corresponds to the state where the total rate of photogeneration in the solar cell equals that of recombination so that no current circulates through the circuit. The relative change of open-circuit voltage with temperature, β_{Voc}, is thus an indication of the temperature dependence of the generation-recombination balance. The photogeneration rate is a function of incident spectrum, concentration, reflection, transmission and parasitic absorption.[15] The recombination rate depends on the type and magnitude of the recombination processes that happen in the device (radiative, SRH, Auger, surface, shunts).

The effect of temperature on PV conversion is discussed in Green's first textbook (Green 1982). The single-diode model (2.17) is used with a general expression of the diode saturation current density (2.18) and the temperature dependence of

[15]Parasitic absorption corresponds to the absorption by the lattice atoms or by free carriers of potentially useful photons, i.e. with $E \geq E_g$. These absorption mechanisms prevent the creation of electron-hole pairs and result in the generation of heat (see Sect. 3.2).

short-circuit current is neglected to express the temperature sensitivity of open-circuit voltage mainly as a function of bandgap and open-circuit voltage (2.19). The corresponding equations read:

$$J \approx J_{sc} - J_0 \exp\left(\frac{qV}{kT_c}\right) \tag{2.17}$$

$$J_0 = A\,T_c^{\gamma} \exp\left(-\frac{E_{g0}}{kT_c}\right) \tag{2.18}$$

$$\frac{d\,V_{oc}}{dT_c} = -\frac{\frac{E_{g0}}{q} - V_{oc} + \gamma\frac{kT_c}{q}}{T_c} \tag{2.19}$$

$$E_{g0} = E_g - T_c\frac{dE_g}{dT_c} \tag{2.20}$$

where A is a factor independent of temperature, E_{g0} is the bandgap of the semiconductor linearly extrapolated to 0 K, and γ includes the temperature dependences of several parameters which determine the diode saturation current density J_0.

Appendix 2 describes the origin of this expression of the diode saturation current, (2.18), that is used with different semi-empirical parameters in many works dealing with temperature coefficients (Fan 1986; Landis 1994; Yoon and Garboushian 1994; Friedman 1996). In short, the γ term accounts for the temperature dependences of several parameters such as the number of effectively available states in the conduction and valence bands, carrier mobilities and lifetimes (which depend on the dominant recombination mechanisms), etc. An interesting discussion on the physical meaning of the parameter J_0 can be found in Cuevas (2014).

Equation (2.19) shows the important relation between the open-circuit voltage and its temperature sensitivity. It predicts an approximately linear temperature dependence of the open-circuit voltage and shows that the temperature coefficient of V_{oc} should decrease as V_{oc} increases. In Green et al. (1982), it is mentioned that "as the open-circuit voltage of silicon solar cells continues to improve, one resulting advantage, not widely appreciated, is reduced temperature sensitivity of device performance". This is an important understanding which is verified experimentally as illustrated in Fig. 2.13 for c-Si solar cells. Note that it is generally valid for all kinds of PV devices.

The parameter γ in (2.19) is not directly related to device physics, which can lead to systematic errors in modeling as demonstrated in Green (2003). In this article, a general analysis of the temperature behavior of a PV cell based on internal device physics is proposed. The temperature dependence of open-circuit voltage is shown to be mainly driven by the recombination mechanisms which depend on the excess carrier concentrations. As a matter of fact, Shockley-Read-Hall, Auger and radiative recombination rates can be written as a function of the product of the electron and hole carrier concentrations (*np*). This *np* product depends on the intrinsic carrier

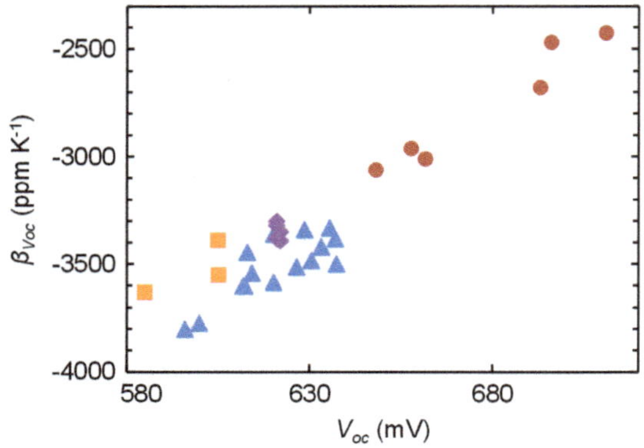

Fig. 2.13 Experimental temperature coefficient of the open-circuit voltage of c-Si solar cells as a function of open-circuit voltage. *Data sources* Triangles from Ponce-Alcántara et al. (2014), squares from Xiao et al. (2014), diamonds from Tanay et al. (2011) and circles from Green et al. (1982, 1985 and Zhao et al. (1994)

concentration, n_i, and on the splitting of the quasi-Fermi levels, E_{Fp}–E_{Fn}, which is related to the potential, V, in the cell:

$$np = n_i^2 \exp\left[\frac{q\,(E_{Fp} - E_{Fn})}{k\,T_c}\right]. \tag{2.21}$$

For convenience, the term ξ, closely related to the np product, is introduced:

$$\xi = np\ \exp\left(-\frac{E_g}{k\,T_c}\right)/n_i^2. \tag{2.22}$$

Regardless of the recombination mechanisms, the output current of a cell (I) can be written as the photogenerated current, I_L, minus the sum of the different recombination processes:

$$I = I_L(V, T_c) - \sum_{i=1}^{N} A_i(V) T_c^{\gamma_i} f_i(\xi_i) \tag{2.23}$$

where the terms $A(V)$ and T^{γ} accommodate for second-order additional dependences on temperature and voltage which can originate from parameters such as diffusion lengths, surface recombination velocities, depletion region width, etc. Using (2.23) in open-circuit configuration and differentiating gives (Green 2003):

$$\frac{dV_{oc}}{dT_c} = -\frac{\left(\frac{\langle E_{g0}\rangle}{q} - V_{oc} + \frac{kT_c}{q}\left\langle \gamma \frac{f}{\xi}\frac{d\xi}{df}\right\rangle - \frac{kT_c}{qI_L}\frac{dI_L}{dT}\left\langle \frac{f}{\xi}\frac{d\xi}{df}\right\rangle\right)}{T_c\left(1 + \frac{kT_c}{q}\left\langle \frac{1}{A}\frac{dA}{dV}\frac{f}{\xi}\frac{d\xi}{df}\right\rangle - \frac{kT_c}{qI_L}\frac{\partial I_L}{\partial V}\left\langle \frac{f}{\xi}\frac{d\xi}{df}\right\rangle\right)} \tag{2.24}$$

where <G> is the weighted value of the parameter of interest defined as:

$$\langle G\rangle = \frac{\sum_{i=1}^{N} G_i A_i T_c^{\gamma_i} \xi_i \left(df_i/d\xi_i\right)}{\sum_{i=1}^{N} A_i T_c^{\gamma_i} \xi_i \left(df_i/d\xi_i\right)}. \tag{2.25}$$

Neglecting the second-order terms in (2.24), the temperature sensitivity of open-circuit voltage can then be written simply as:

$$\frac{dV_{oc}}{dT_c} = -\frac{\left(\frac{\langle E_{g0}\rangle}{q} - V_{oc} + \frac{kT_c}{q}\left\langle \gamma \frac{f}{\xi}\frac{d\xi}{df}\right\rangle\right)}{T_c}. \tag{2.26}$$

This expression is similar to (2.19) but the term $<\gamma \frac{f}{\xi}\frac{d\xi}{df}>$ provides a direct link to the underlying physics. The expressions (2.24) and (2.26) can be applied to any kind of PV cell as they are derived from perfectly general arguments but they require the detailed knowledge of the dominant recombination mechanisms.

It is possible to derive another similar expression for the temperature sensitivity of open-circuit voltage using the concept of external radiative efficiency (ERE). The ERE of a PV cell is similar to the external quantum efficiency (EQE) of a light-emitting diode (LED). It is defined as "the fraction of the total dark current recombination in the device that results in radiative emission from the device" (Green 2012). Note that the ERE is different from the ratio of radiative recombination rate on total recombination rate, which is the internal radiative efficiency (IRE), because of photon recycling. Indeed, only a fraction of the photons emitted through radiative recombinations actually exits the cell (see for example Fig. 3.2). Using the definition of the ERE, the output current and the open-circuit voltage of the cell can be written as (Dupré et al. 2015b):

$$\begin{aligned} J &= X\,J_{sc,1sun} - \frac{1}{ERE} J_{0,rad}\left(\exp\left(\frac{qV}{kT_c}\right) - 1\right) \\ &\approx X\,J_{sc,1sun} - \frac{1}{ERE} J_{0,rad}\exp\left(\frac{qV}{kT_c}\right) \end{aligned} \tag{2.27}$$

$$V_{oc} = \frac{kT_c}{q}\ln\left(\frac{X\,J_{sc,1sun}}{(1/ERE_{oc})\,J_{0,rad}}\right) = V_{oc,1sun} + \frac{kT_c}{q}\left(\ln(ERE_{oc}) + \ln(X)\right) \tag{2.28}$$

where $J_{sc,1sun}(T_c) = q \int_{E_g(T_c)}^{\infty} PFD\,(E)\,dE$, ERE_{oc} is the ERE at open circuit, X is the concentration factor and $J_{0,rad}$ is the dark current density in the radiative limit. Transport resistances are neglected. Writing $X\,J_{sc,1sun}$ instead of J_{sc} means that a linear behavior of J_{sc} with the concentration factor X is assumed. This assumption can fail if an important loss mechanism in short circuit has a significant intensity dependence. Nonetheless, (2.28) shows that increasing the concentration factor generally improves open-circuit voltage and in turn improves its temperature coefficient. A specific discussion about concentrator PV devices is presented in Sect. 4.7.

Similarly to concentration, the effect of non-ideal absorption could easily be included in (2.28) and (2.30) through the collection fraction f_c (see Sect. 2.3.2). However, in practice, its impact on the temperature coefficient of open-circuit voltage is limited (e.g. a collection fraction increase from 60 to 80% causes only 2.6% change in β_{Voc}).

The dark current density in the radiative limit can be written as[16]:

$$J_{0,rad} \approx q \frac{2\,\Omega_{emit}}{c^2\,h^3} k\,T_c\,E_g^2 \exp\left(-\frac{E_g}{k\,T_c}\right). \tag{2.29}$$

Assuming a linear variation of E_g on the temperature range of interest, $E_g = E_{g0} + T_c \frac{dE_g}{dT_c}$, and differentiating (2.28), the temperature dependence of V_{oc} reads (Dupré et al. 2015b):

$$\frac{dV_{oc}}{dT_c} = -\frac{\frac{E_{g0}}{q} - V_{oc} + \gamma \frac{kT_c}{q}}{T_c} \tag{2.30}$$

with:

$$\gamma = 1 - \frac{d\ln ERE_{oc}}{d\ln T_c} + \left(2\frac{d\ln E_g}{d\ln T_c} - \frac{d\ln J_{sc,1sun}}{d\ln T_c}\right). \tag{2.31}$$

The expression (2.30) is the same as (2.19) but the coefficient γ is explicitly quantified in (2.31). It corresponds to the temperature sensitivity of the mechanisms determining V_{oc}, similarly to the term $\left\langle \gamma \frac{f}{\xi} \frac{d\xi}{df} \right\rangle$ in (2.26). One advantage of using the concept of external radiative efficiency, is that it enables to compare solar cells of both the same and completely different technologies (Green 2012). Making the rough approximations of neglecting the temperature dependences of ERE and bandgap, Fig. 2.14 shows temperature coefficients as a function of cell bandgap for different external radiative efficiencies. This gives an idea of the evolution of temperature coefficients as cells improve towards the radiative limit. For example, the crosses in Fig. 2.14 correspond to experimental values of β_{Voc} for record-efficiency crystalline silicon cells over the years. Note that the relation between *ERE* and β_{Voc} is not so straightforward because ERE and bandgap temperature dependences are not always negligible. Figure 2.15 illustrates that dE_g/dT

[16]This expression is derived from Planck's equation using Boltzmann's approximation and neglecting $2E_gkT + 2k^2T^2$ in front of E_g^2 in the integration by parts (Dupré 2015).

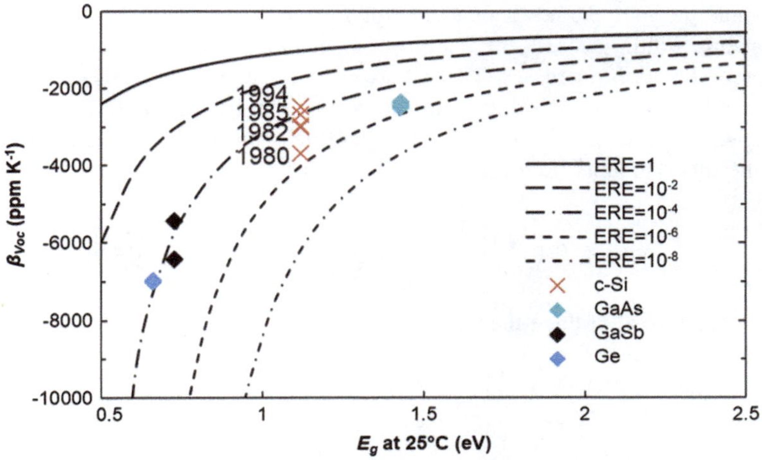

Fig. 2.14 Temperature coefficient of open-circuit voltage as a function of bandgap for different values of the external radiative efficiency (dE_g/dT and $dERE/dT$ are neglected in the calculation). The crosses are experimental values of β_{Voc} for some record-efficiency c-Si cells over the years (Green et al. 1982, 1985; Zhao et al. 1994). The diamonds are experimental values of β_{Voc} for GaAs, GaSb and Ge cells (Siefer et al. 2005). Figure adapted from Dupré et al. (2015b)

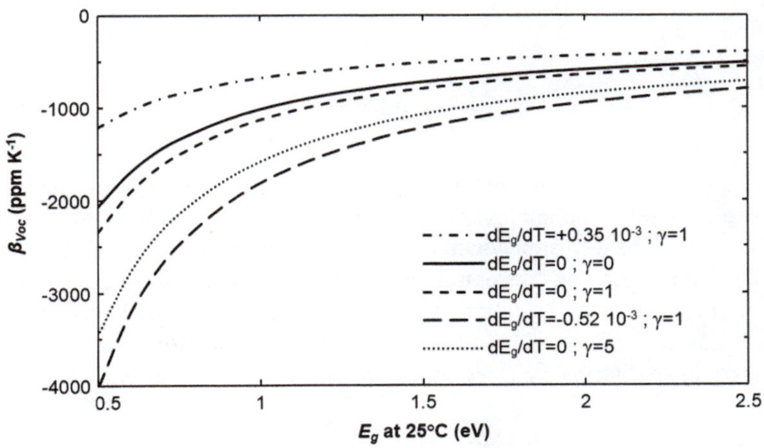

Fig. 2.15 Temperature coefficient of open-circuit voltage as a function of bandgap for $ERE = 1$ and different values of dE_g/dT and γ. Figure from Dupré et al. (2015b)

and γ can have a significant impact on β_{Voc}. The value of γ depends on $dERE/dT$ which is function of the dominant recombination mechanisms. Consequently, the knowledge of V_{oc} or ERE at 25 °C alone is not sufficient to accurately predict β_{Voc}.

The absolute value of the terms in parentheses in (2.31) is usually less than 0.5 (Dupré 2015). Thus, it is $1 - d\ln ERE_{oc}/d\ln T_c$ that plays a major role in the value of γ. Since the temperature dependence of ERE is function of the recombination mechanisms within the cell, the parameter γ contains an information on the dominant recombination processes. This idea can be illustrated by analyzing two situations where the saturation current is dominated by different recombination mechanisms (Siefer and Bett 2012). The starting point is the single-diode model using the diode ideality factor, n, and the dark saturation current density J_0:

$$J = J_{sc} - J_0\left(\exp\left(\frac{qV}{nkT_c}\right) - 1\right) \approx J_{sc} - J_0 \exp\left(\frac{qV}{nkT_c}\right). \tag{2.32}$$

The open-circuit voltage then reads:

$$V_{oc} \approx \frac{nkT_c}{q} \ln\left(\frac{J_{sc}}{J_0}\right). \tag{2.33}$$

- Scenario 1: the dominant recombination mechanism is recombination through defects (Shockley-Read-Hall) in the bulk regions and at the surfaces. In this case, recombination rates are only limited by minority carrier concentrations and the ideality factor is equal to one. The dark saturation current density can be written as (Green 1982):

$$J_{01} = q\left(\frac{D_e}{L_e}\frac{1}{N_A}F_P + \frac{D_h}{L_h}\frac{1}{N_D}F_N\right) n_i^2 \tag{2.34}$$

where F_P and F_N are factors that account for the finite dimensions of the P and N regions respectively (and are thus function of the surface recombination velocities). $D_{e,h}$ and $L_{e,h}$ are the diffusion coefficients and the diffusion lengths of electrons and holes. $N_{A,D}$ are the doping concentration of acceptors and donors respectively and n_i is the intrinsic carrier concentration. Neglecting the temperature dependence of $D_e/L_e\, F_P$ and $D_h/L_h\, F_N$, the temperature dependence of J_0 comes only from n_i^2. The square of the intrinsic carrier concentration reads (Sze and Ng 1981):

$$\begin{aligned} n_i^2 &= N_c N_v \exp(-E_g/kT_c) \\ &= 4\,(2\pi k\, T_c/h^2)^3 (m_e\, m_h^*)^{3/2} \exp(-E_g/kT_c). \end{aligned} \tag{2.35}$$

Neglecting the temperature dependence of the electron and hole effective masses, m_e^* and m_h^*, it follows that:

$$n_i^2(T_c) \propto T_c^3 \exp(-E_g(T_c)/kT_c). \tag{2.36}$$

Thus:

$$J_{01}(T_c) \propto T_c^3 \exp\left(-\frac{E_g(T_c)}{kT_c}\right). \tag{2.37}$$

Using (2.37) in the derivation of (2.33) gives:

$$\frac{dV_{oc}}{dT_c} = -\frac{\frac{E_{g0}}{q} - V_{oc} + \frac{kT_c}{q}\left(3 - \frac{d\ln J_{sc}}{d\ln T_c}\right)}{T_c}. \tag{2.38}$$

Equation (2.38) corresponds to (2.30) with $\gamma = 3 - \frac{d\ln J_{sc}}{d\ln T_c} \approx 3$.

- Scenario 2: the dominant recombination mechanism is recombination through defects (Shockley-Read-Hall) in the depletion zone. In this case, the recombination rate is limited by both minority and majority carrier concentrations and the ideality factor is equal to two. The dark saturation current density can be written as (Siefer and Bett 2012):

$$J_{02} = \frac{n_i\, W_d\, k\, T_c}{2(V_d - V)\tau_{n0}} \tag{2.39}$$

where W_d is the width of the depletion zone, V_d is the diffusion voltage also known as built-in voltage and τ_{n0} is the minimum electron lifetime. Neglecting the temperature dependence of these parameters and using (2.36) gives:

$$J_{02} \propto T_c^{2.5} \exp\left(-\frac{E_g(T_c)}{2kT_c}\right). \tag{2.40}$$

Using (2.40) in the derivation of (2.33) results in:

$$\frac{dV_{oc}}{dT_c} = -\frac{\frac{E_{g0}}{q} - V_{oc} + \frac{kT_c}{q}\left(5 - 2\frac{d\ln J_{sc}}{d\ln T_c}\right)}{T_c} \tag{2.41}$$

which corresponds to (2.30) with $\gamma = 5 - 2\frac{d\ln J_{sc}}{d\ln T_c} \approx 5$.

These scenarios illustrate the impact of the dominant recombination mechanisms on γ and correspondingly the information contained in this parameter. Of course, there exist other recombination mechanisms that have different temperature dependences such as Auger or radiative recombination. For example, in the radiative limit (*ERE* = 1), (2.31) gives $\gamma \approx 1.2$ for crystalline silicon. Also, it is possible that, in some cases, the assumptions made in the two scenarios presented above (neglecting the temperature dependence of the carrier lifetime or the surface recombination velocity for example) do not hold, leading to different values for γ.

Besides, in actual devices, several mechanisms with different temperature and voltage dependences often contribute to the global recombination rate. In this case, the J-V characteristic has to be modeled with a double (or more) diode model and it becomes complicated to derive an analytical expression of dV_{oc}/dT_c using the same approach. Instead of using the diode model, Grover et al. propose another formulation of solar cell physics that enables to assess quantitatively the strengths of the different possible recombination pathways from simple J-V measurements on a range of temperatures and intensities (Grover et al. 2013). Quite similarly to the derivation in Green (2003), the recombination rates and the open-circuit voltage are expressed as a function of a parameter ς closely related to the np product:

$$V_{oc} = \frac{kT_c}{q}\ln\left(\frac{np}{n_i^2}\right) = \frac{kT_c}{q}\ln\left(\varsigma^2\right). \tag{2.42}$$

Expressing that the total generation rate in the cell is equal to the recombination rate at open circuit gives:

$$\left(R_0^I + R_0^I\right)\varsigma^2 + R_0^I\varsigma = \int_W G_x dx = G_{avg}W \tag{2.43}$$

where G_{avg} is the total generation rate divided by the cell thickness, W, and R_0^b, R_0^d and R_0^I correspond to the strengths of the recombination mechanisms independent of bias conditions (respectively in the bulk, the depletion region and at the interfaces). These parameters R_0 carry the same physical meaning as the parameters usually denoted J_0 in the diode models.

Solving (2.43) gives the following expression of the open-circuit voltage:

$$V_{oc} = 2\frac{kT_c}{q}\ln\left[k_1\left(\sqrt{1 + k_2G_{avg}} - 1\right)\right] \tag{2.44}$$

with:

$$k_1 = \frac{R_0^d}{2\left(R_0^I + R_0^b\right)}; \quad k_2 = 4W\frac{\left(R_0^I + R_0^b\right)}{\left(R_0^d\right)^2}. \tag{2.45}$$

This set of equations contains the temperature and intensity dependence of V_{oc} and shows the dependence of V_{oc} on the different recombination processes via R_0^b, R_0^d and R_0^I. Thus, it is possible to use this theoretical framework to identify and quantify different recombination pathways in PV devices using simple electrical measurements (i.e. temperature and intensity dependent J-V measurements). With this method, it is possible to extract valuable information on device physics as demonstrated for silicon heterojunctions, CIGS, C_uO_2 and GaAs solar cells in Grover et al. (2013), Li et al. (2014a, b), Grover and Li (2015).

While the analytical method presented above can provide useful insights into device physics, it often requires some prior knowledge of the device considered. Another approach consists in fitting temperature and intensity dependent V_{oc} measurements to a complete 1-D drift-diffusion model (Brandt et al. 2016). This type of numerical modelling associated with simple electrical measurements may be an interesting alternative to longer and more costly spectroscopy studies in the ongoing search for new materials for PV applications.

2.3.2 Short-Circuit Current Temperature Sensitivity

The short-circuit current density, J_{sc}, generally increases slightly with temperature. This is mainly due to the fact that most semiconductor bandgaps decrease with temperature (see Fig. 2.7). The short-circuit current density depends on the number of photons able to create an electron-hole pair, i.e. with $E > E_g$, and on the loss mechanisms in short circuit such as parasitic absorption or surface recombination for example. Thus, J_{sc} can be expressed as the product of an ideal short-circuit current density $J_{sc,ideal}$ and a collection fraction f_c (Green 2003):

$$J_{sc} = J_{sc,ideal} f_c. \tag{2.46}$$

The ideal short-circuit current density is determined by the photon flux density (PFD) of the incident radiation and the cell bandgap:

$$J_{sc,ideal}(T_c) = q \int_{E_g(T_c)}^{\infty} PFD\,(E)\,dE. \tag{2.47}$$

The temperature dependence of $J_{sc,ideal}$ depends on the temperature dependence of the bandgap. If the bandgap decreases when temperature rises, more electron-hole pairs can potentially be photogenerated and $dJ_{sc,ideal}/dT_c$ is positive. For semiconductors such as perovskites whose bandgaps increase with temperature, the temperature dependence of $J_{sc,ideal}$ is negative.

The temperature coefficient of the short-circuit current can be written as (Green 2003):

$$\beta_{J_{sc}} = \frac{1}{J_{sc}} \frac{dJ_{sc}}{dT_c} = \frac{1}{J_{sc,ideal}} \frac{dJ_{sc,ideal}}{dE_g} \frac{dE_g}{dT_c} + \frac{1}{f_c} \frac{df_c}{dT_c} \tag{2.48}$$

The collection fraction f_c is the fraction of potentially useful photons ($E \geq E_g$) that excites a carrier that gets collected in short circuit. It accounts for optical losses such as reflection, insufficient or parasitic absorption as well as electrical losses such as bulk or surface recombination that happen in short-circuit condition. It is closely related to the external quantum efficiency (EQE) and can be expressed as:

$$f_c(T_c) = \frac{J_{sc}(T_c)}{J_{sc,ideal}(T_c)} = \frac{q\int_0^{\infty} EQE(E)\,PFD\,(E)\,dE}{q\int_{E_g(T_c)}^{\infty} PFD\,(E)\,dE} \tag{2.49}$$

The expression (2.48) of the temperature coefficient of J_{sc} is interesting in that it separates the influence of intrinsic material properties from that of device parameters. The first summand is a function of the solar cell material (its bandgap and its bandgap temperature sensitivity) and the intensity and spectrum of the irradiation. The second summand is related to the cell light trapping ability and collection efficiency. It can be derived from external quantum efficiency (EQE) measurements on a range of temperatures.

The first summand in (2.48) indicates that β_{Jsc} depends on the incident spectral intensity at wavelengths near the bandgap. Figure 2.16 shows the photon flux density of the reference AM1.5 spectrum (IEC 2008) and that of a solar simulator (Emery 2008) together with the bandgaps of several semiconductors at 0 and 100 °C. While the average intensity of solar simulators is close to that of the reference spectrum, the photon flux densities can be quite different as illustrated in Fig. 2.16. This explains the scattering of β_{Jsc} values in the literature (Landis 1994). Even for indoor measurements, the spectral intensity distributions vary between different solar simulators (even of the same type). This stresses the complexity of accurately predicting β_{Jsc} under real operating conditions, where the incident spectrum changes over time.

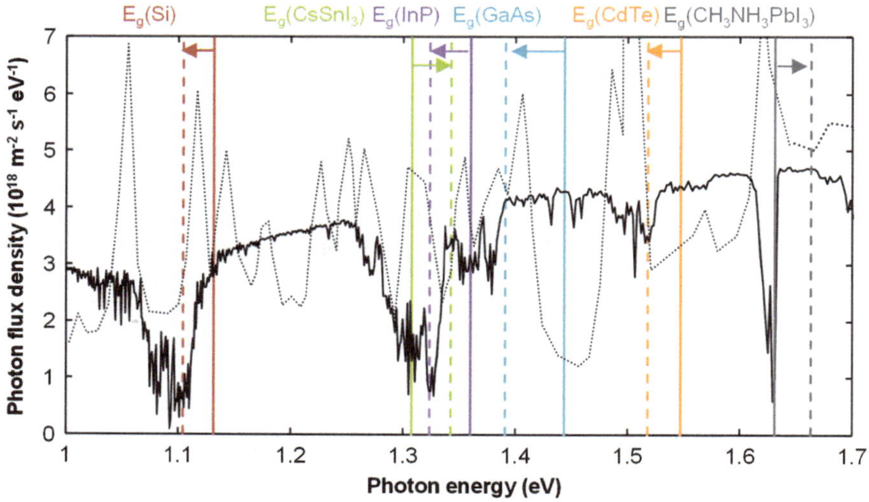

Fig. 2.16 Photon flux density of the AM1.5 reference spectrum (*solid line*), of a solar simulator (Emery 2008) (*dotted line*). Bandgaps of different semiconductors at 0 and 100 °C (shown by *solid* and *dashed lines* respectively); the *arrows* show the direction of increasing temperature. Figure adapted from Dupré et al. (2015b)

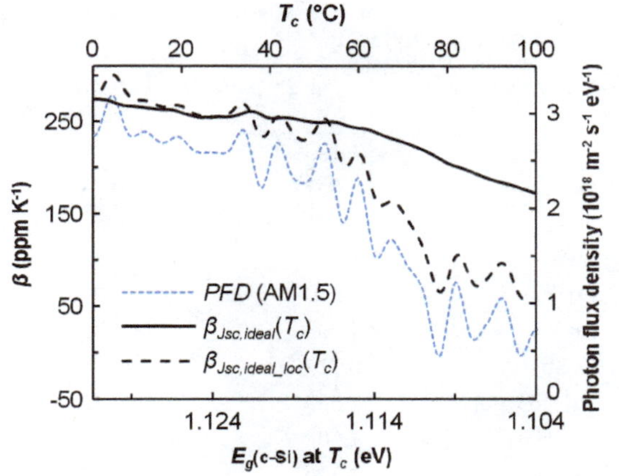

Fig. 2.17 Global (*solid line*) and local (*dashed line*) temperature coefficients of the short-circuit current density of a c-Si cell with ideal light collection ($f_c = 1$). Photon flux density of the AM1.5 reference spectrum (*dotted line*). Figure adapted from Dupré et al. (2015b)

Figure 2.16 shows that some semiconductor bandgaps lie near important fluctuations in the AM1.5 photon flux density. This induces non-linearity in the temperature dependence of the ideal short-circuit current density. Figure 2.17 illustrates this phenomenon in the case of c-Si. The temperature coefficient of short-circuit current density undergoes a change of 37% between 0 and 100 °C. This variation can be related to the photon flux density fluctuations. To represent the local variations of $J_{sc,ideal}$, a local temperature coefficient ($\beta_{G,loc}$) is defined (in ppm K^{-1}) as:

$$\beta_{G,loc}(T_c) = \frac{10^6}{G(298.15\,\mathrm{K})} \frac{G(T_c+1) - G(T_c-1)}{2} \tag{2.50}$$

Figure 2.17 shows that the local variations of the ideal short-circuit current density are proportional to the local variations of the photon flux density. This illustrates the impact of the photon flux density on the temperature coefficient of the short-circuit current density.

The first summand in (2.48) accounts for the change in the number of photons with $E > E_g$ caused by the temperature dependence of the bandgap. This term explains the difference in β_{Jsc} between cells based on different semiconductors. For example, using dE_g/dT from Table 2.2 and the photon flux density of the AM1.5 spectrum, this term is equal to about 250, 460, 670 and -450 for c-Si, GaAs, CdTe and $CH_3NH_3PbI_3$ respectively.

The temperature coefficients of the short-circuit current density of actual devices can differ from these values because of the second summand in (2.48) which accounts for non-idealities. In general, the collection fraction increases with temperature and $1/f_c \times df_c/dT_c$ is positive. Note that this term is usually larger for indirect bandgap materials because phonons play an important role in their interband absorption. The positive temperature dependence of interband absorption results in an interesting trend for c-Si cells: β_{Jsc} tends to decrease when J_{sc} increases (Fig. 2.18). One possible explanation is that, regardless of the loss mechanism, the

Table 2.2 Bandgaps of various semiconductors at room temperature and values of dE_g/dT from linear fits around 300 K

Semiconductor	E_g @ 25 °C (eV)	dE_g/dT (meV K^{-1})	Source for $E_g(T)$
c-Si	1.125	−0.27	Green (1990)
Ge	0.661	−0.38	Varshni (1967)
GaAs	1.431	−0.51	
InP	1.351	−0.36	
CdTe	1.540	−0.29	Singh and Ravindra (2012)
CdS	2.503	−0.36	
GaSb	0.727	−0.38	Vurgaftman et al. (2001)
$CH_3NH_3PbI_3$	1.639	0.32	Jiang et al. (2016)
$CH_3NH_3PbI_{3-x}Cl_x$	1.608	0.30	Wu et al. (2014)
$CsSnI_3$	1.316	0.35	Yu et al. (2011)

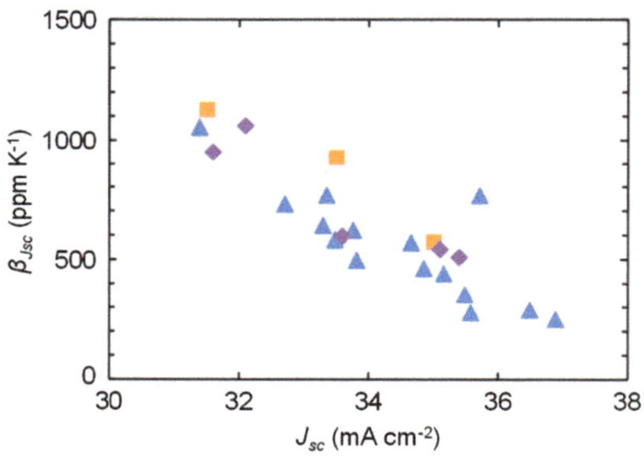

Fig. 2.18 Experimental temperature coefficients of short-circuit current density as a function of short-circuit current density. *Data sources*: *Triangles* from Ponce-Alcántara et al. (2014), *squares* from Xiao et al. (2014) and *diamonds* from Tanay et al. (2011)

cells that collect less electron-hole pairs at room temperature benefit relatively more from the increased absorption with temperature. This effect is illustrated in Fig. 2.19 where the gain in EQE when the temperature increases from 25 to 40 °C is highlighted for two c-Si cells with different short-circuit current densities. Because the cell with a low EQE collects less of the potentially useful photons at room temperature, it undergoes a larger boost in EQE, and thus in collection fraction, when the temperature rises. Note that the trend apparent in Fig. 2.18 is not a general rule and devices with different light trapping structures can present different trends (Dupré 2015).

Similarly to the temperature coefficient of V_{oc}, that of J_{sc} contains information on some of the loss mechanisms in the solar cell. To obtain some insight into the physics of carrier collection, it is interesting to analyse the variations with temperature of the external quantum efficiency. In Philipps et al. (2011), a numerical model including the temperature dependences of material parameters (e.g. bandgap)

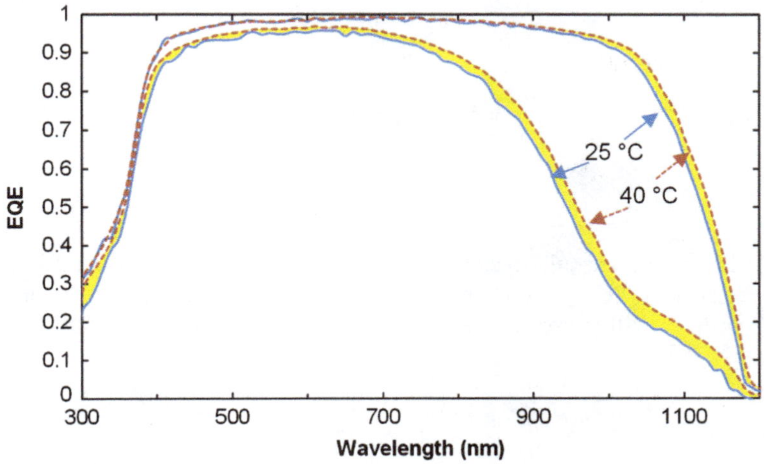

Fig. 2.19 EQEs at 25 and 40 °C (*solid* and *dashed lines* respectively) of two c-Si cells. The variations of EQE between 25 and 40 °C are highlighted. Figure created with data from Dupré et al. (2015a)

and physical processes (e.g. recombination rates) is used to extract information from measurements of the EQE of GaAs cells over a range of temperature. The analysis enables a better understanding of the thermal behavior of these cells. Another example of the use of temperature dependent EQE measurements can be found in Søndenå et al. (2015). The EQEs from compensated and non-compensated polysilicon cells are compared and the different changes with temperature are associated with differences in minority carrier lifetimes and mobilities of the bulk materials.

2.3.3 Fill Factor Temperature Sensitivity

The fill factor (*FF*) relates the maximum power that can be extracted from a cell (P_{MPP}) to the product of its open-circuit voltage (V_{oc}) and short-circuit current (I_{sc}). It indicates the minimal "cost" of extracting the photogenerated charges from the cell into the circuit and corresponds to the optimal current/voltage trade-off. This optimum depends on the generation-recombination balance (similarly to V_{oc}) but also on the transport losses associated with the current flow through the circuit at the maximum power point (MPP).

When both series and shunt resistances have a negligible impact on cell performance, the fill factor can be accurately expressed as (Green 1981):

$$FF_0 = \frac{v_{oc} - \ln(v_{oc} + 0.72)}{v_{oc} + 1} \tag{2.51}$$

where v_{oc} is the normalized open-circuit voltage defined as:

$$v_{oc} = \frac{n\,k\,T_c}{q} V_{oc} \tag{2.52}$$

where n is the diode ideality factor. Assuming that the ideality factor and series and shunt resistances do not vary significantly with temperature, the temperature coefficient of the fill factor can be approximated as (Green et al. 1982):

$$\beta_{FF_0} = \frac{1}{FF_0}\frac{dFF_0}{dT_c} \approx (1 - 1.02\,FF_0)\left(\frac{1}{V_{oc}}\frac{dV_{oc}}{dT_c} - \frac{1}{T_c}\right) \tag{2.53}$$

This expression highlights the relation between β_{FF} and β_{Voc}, which results from the fact that the fill factor also depends upon the generation-recombination balance in the cell. Note that (2.53) neglects the effect of parasitic transport losses.

When the series resistance, R_s, is important and the shunt resistance is so large as to have a negligible effect, the fill factor is better described by Green (1981):

$$FF_s = FF_0\left(1 - \frac{R_s}{V_{oc}/I_{sc}}\right). \tag{2.54}$$

The temperature coefficient of the fill factor becomes (Zhao et al. 1994):

$$\beta_{FF} = \frac{1}{FF}\frac{dFF}{dT_c} \approx (1 - 1.02\,FF_0)\left(\frac{1}{V_{oc}}\frac{dV_{oc}}{dT_c} - \frac{1}{T_c}\right) - \frac{R_s}{V_{oc}/I_{sc} - R_s}\left(\frac{1}{R_s}\frac{dR_s}{dT_c}\right). \tag{2.55}$$

In this expression, certain transport losses appear in the second summand. This term can be significant especially in early, non-optimized, devices. Even in the case of a record efficiency PERC cell, this second summand is 303 ppm K^{-1} which corresponds to 25% of the total temperature coefficient of the fill factor (Zhao et al. 1994). While (2.55) takes into account series resistance, it has been observed that it does not accurately predict the fill factor temperature sensitivity of certain solar cells (Dupré et al. 2015a). The discrepancies between the measurements and the values of β_{FF} predicted by (2.55) may be due to variations with temperature of the ideality factor and the shunt resistance which are neglected in the derivation of (2.55). These effects are particularly important in certain devices. For example, in certain amorphous silicon and nanocrystalline dye cells, the fill factor increases with temperature (which is the opposite of the usual trend) due to decreasing resistance effects or increasing "mobility-lifetime" products (Green 2003; Riesen et al. 2016). It is also interesting to note that the impact of series resistances generally increases with irradiance because of the augmentation of current circulating through the cell (Helmers et al. 2013). Similarly, the effect of shunt resistance is more important at low illumination levels. Additionally, certain limiting transport mechanisms cause important non-linearities in β_{FF} which result in non-linear β_η (see Sects. 4.2 and 4.4).

An interesting example of the relation between the limiting transport mechanisms and the temperature coefficient of the fill factor in CIGS modules is reported in Deceglie et al. (2014). The temperature coefficient of the fill factor of CIGS modules is found to be increased after light soaking. It is suggested that this is due to a light induced reduction in the conduction band offset between the buffer and the absorber. Since thermionic emission is the limiting charge transfer mechanism across this barrier and it increases with temperature (Sze and Ng 1981), this metastable change could explain the observed variation of β_{FF}. In other words: the decrease of interface resistance with temperature depends upon the height of the barrier that carriers have to go through. Light soaking modifies this barrier height and thus impacts the temperature coefficient of the fill factor. This understanding is important because it is the final light soaked temperature coefficients that need to be known in order to predict field performances.

Another example of the impact of transport barriers on the temperature sensitivity of the fill factor is highlighted in Seif et al. (2014) for silicon heterojunction (SHJ) cells. In this work, the intrinsic amorphous silicon layer at the front is replaced by an amorphous silicon oxide layer in order to reduce optical losses. However, at room temperature, the gain in short-circuit current is more than compensated by a loss in fill factor. This loss in fill factor is attributed to a valence band offset impeding the collection of holes. The effect of this transport barrier decreases with temperature because, unlike standard devices which operate in a carrier-diffusion-limited regime, the limiting charge transfer mechanism here is thermionic emission of the carriers across the barrier. Once again, this transport barrier whose effect decreases with temperature results in a more favorable temperature coefficient of the fill factor. This effect of transport barriers on β_{FF} are also reported in SHJ cells when using thicker amorphous silicon layers (Taguchi et al. 2008) or a MoOx hole selective contact (Battaglia et al. 2014). Additionally, the results in Seif et al. (2016) suggest that the thermal behavior of SHJ cells is impacted not only by transport-related phenomena but also by the passivation properties of the interfaces.

In Seif et al. (2014), the advantageous β_{FF} of the device with the amorphous silicon oxide layer results in a better β_η which translates into better performances than the standard device at temperatures above 45 °C. This result demonstrates that there exist opportunities to tune the temperature coefficients of solar cells and optimize their designs at their operating temperature instead of at 25 °C (which is the temperature prescribed in the standard test conditions, see Sect. 3.5).

2.4 Tuning the Temperature Coefficients

While the importance of temperature coefficients for the energy yield of PV devices is well known (it is even advertised by several PV manufacturers), the idea of tuning the temperature coefficients is seldom mentioned in the literature. It is now understood that temperature coefficients generally improve together with open-circuit voltage so that improved temperature sensitivity automatically follows from ameliorations in V_{oc}. This explains, for example, the advantageous temperature coefficients of silicon heterojunction solar cells also known as HIT[17] cells (see Sect. 4.2). Another illustration of the positive impact of open-circuit voltage is the reduced temperature coefficients of PV systems operating under concentration (e.g. Yoon and Garboushian 1994). Note that while PV systems under concentration have reduced temperature sensitivities, the increased incoming irradiance also results in an increased heat generation in the device (see Sect. 4.7). Interestingly, the concept of reduction of the angle of emission should theoretically improves temperature sensitivity (see Sect. 2.2.1) without increasing the incoming power and thus the heat source, making this solution ideal in terms of thermal behavior. However, thermal behavior and temperature coefficients are rarely considered as design parameters of PV devices. Furthermore, as discussed in the previous sections, open-circuit voltage is not the only factor that influences the temperature sensitivities of a PV device performances.

The main driver of the temperature coefficient of the efficiency of a solar cell is its bandgap energy (see Sect. 2.2). This is why, in general, cells made of large bandgap materials such as GaAs or CdTe have better temperature coefficients than that those made of c-Si for example. Also, each semiconductor has a specific bandgap temperature dependence caused by electron-phonon interactions and by thermal expansion of the lattice (see Sect. 2.2.2). Depending on the value of the bandgap at room temperature, the bandgap temperature dependence (dE_g/dT) can affect either positively or negatively the temperature sensitivity of PV conversion efficiency. Figure 2.8 illustrates that, for an AM1.5 illumination, cells with bandgaps larger than ~ 1.5 eV benefit, in terms of temperature coefficient, from negative dE_g/dT while cells with bandgaps lower than ~ 1.5 eV benefit from positive dE_g/dT (such as that of perovskite compounds). Several articles indicate that the bandgap temperature sensitivities of certain quantum dots (QD) change with their size (Olkhovets et al. 1998; Dey et al. 2013). This highlights a potential way of tuning dE_g/dT to optimize the temperature coefficients of QD solar cells. Another way of adjusting dE_g/dT could be via mechanical stress. For example, in ultra-thin micro-transfer printed cells, the thermal expansion of the substrate impacts the

[17]HIT stands for Heterojunction with Intrinsic Thin layers. This kind of silicon heterojunction solar cells consists of a layer of c-Si, which serves as the absorber, sandwiched between two very thin layers (few nanometers) of amorphous silicon with different doping, which serve as selective membranes. These layers are separated by very thin layers of excellent quality amorphous silicon which passivate the interfaces and enable high open-circuit voltages.

thermal expansion of the lattice of the cell active layer and thus modifies dE_g/dT (Menard et al. 2012). Other ways to modify a semiconductor bandgap include doping at high concentration [effect known as bandgap narrowing (Sze and Ng 1981)] and applying an hydrostatic pressure (Goni et al. 1990).

The previous sections have highlighted that temperature coefficients of solar cells depend on the temperature dependences of all the limiting loss mechanisms. This suggests that the temperature coefficients of a device can be tuned by modifying its design. For example, in the case of space applications, it is suggested in Landis (1994) that it could be worth improving the open-circuit voltage at the expense of performances at 25 °C (by increasing base doping of the cell for example) to reduce the temperature sensitivity of the devices. Ideally, this could lower the temperature-related losses and lead to better performances in real operating conditions. Another example can be found in Seif et al. (2015) where it is proposed that the temperature coefficient of the fill factor can be tuned by adding layers that impede carrier transport at room temperature. In Riesen et al. (2016), the effect of several design parameters of a-Si:H solar cells (i-layer thickness, H_2/SiH_4 flow ratio, etc.) on J-V parameters and their temperature coefficients is studied experimentally. The results highlight the impact of cell design on its thermal behavior and thus on energy yield. This work also shows that an optimum cell design depends on the range of operating temperature which originates from the operating conditions (climate of the location, mounting configuration, etc.).

The potential of tuning the temperature coefficients in order to optimize a solar cell so that it performs better at its actual operating temperature is demonstrated in Seif et al. (2014). In this work, a new cell design, while less efficient in the standard test conditions, is shown to outperform the standard design above a certain temperature thanks to an advantageous temperature coefficient. This means that the device with the highest efficiency in the STC do not necessarily provide the highest energy output. This idea is illustrated in Fig. 2.20 with the results mentioned above (Seif et al. 2014). The efficiency of the SHJ cell with the original feature (an amorphous silicon oxide layer) is plotted in green and the efficiency of the standard SHJ cell is plotted in orange. As mentioned, the modified cell design underperforms at 25 °C but benefits from an advantageous temperature coefficient which translates into better performances than the standard device at temperatures above 45 °C. The bar charts in Fig. 2.20 represent the distribution of annual energy production as a function of module temperature in Norway (blue bars) and in the US desert Southwest (red bars). It appears that, in the US desert Southwest, more than 50% of the annual energy production is generated when the module temperature is above 45 °C. This means that the modified SHJ cell design would have a better energy yield in this location than a standard SHJ cell (but not if it was installed in Norway). This demonstrates that there exist opportunities to tune the temperature coefficients of solar cells and optimize their designs as a function of their operating temperature instead of at 25 °C. This concept of thermal design of PV devices is discussed further in Chap. 3, Sect. 3.5.

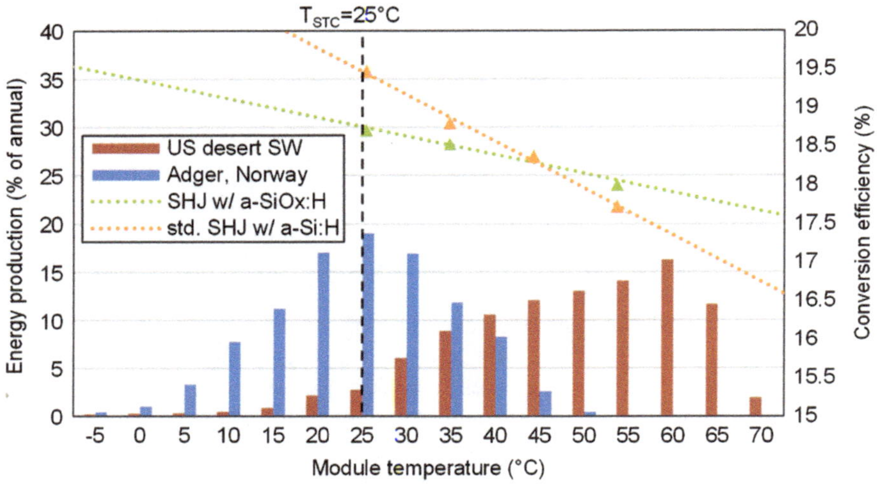

Fig. 2.20 Distribution of annual energy production as a function of module temperature in Norway (*blue bars*) and in the US desert SouthWest (*red bars*). Conversion efficiency as a function of device temperature of a standard SHJ cell (*orange dots*) and a SHJ cell with an amorphous silicon oxide windows layer (*green dots*). Data from Strevel et al. (2012), Seif et al. (2014) and data provided by the Energy Materials Group at the University of Agder (Norway)

2.5 Conclusion

The efficiency of photovoltaic conversion is fundamentally dependent upon the converter temperature. This temperature dependence is negative and its magnitude depends upon the bandgap of the PV device. In general, solar cells made of larger bandgap semiconductors have lower temperature sensitivities. In addition, bandgap temperature dependence, which varies among different materials, modifies the temperature sensitivities as a function of the radiative spectrum incident upon the device.

To these intrinsic effects, add several extrinsic effects dues to non-idealities in existing devices. Each loss mechanism in a solar cell has a specific voltage and temperature dependence. Therefore, the temperature dependence of certain solar cell parameters (V_{oc}, J_{sc}, FF) which correspond to different bias configurations (open circuit, short circuit, MPP) can provide some information about the dominating loss mechanisms.

In open-circuit configuration, the total recombination rate is equal to the total generation rate. Thus, the open-circuit voltage is a function of the different recombination processes that happen in the device. The temperature dependence of open-circuit voltage can be used to identify and quantify the dominant recombination pathways by their functional dependences.

In short-circuit configuration, the current depends on the photogeneration of electron-hole pairs and on the efficiency at which they are collected. The photogeneration depends on the incident spectrum and the bandgap as well as on the optical design of the device. The temperature dependence of the photogeneration mainly comes from that of the bandgap of the absorbing material. On the other hand, the temperature dependence of the collection fraction gives some indication about the loss mechanisms responsible for the incomplete absorption or collection of the available photons. In general, this term is larger for indirect bandgap materials such as c-Si because their interband absorption coefficients increase with temperature.

The fill factor, which links P_{MPP} to V_{oc} and J_{sc}, depends on the generation-recombination balance at MPP which is usually related to the generation-recombination balance in open circuit, hence a relation between β_{FF} and β_{Voc}. At the maximum power point, there is a current flowing through the device so the fill factor is also a function of the limiting transport mechanisms.

The temperature coefficients of PV devices are function of a large number of parameters that can be tuned by design. There exist several opportunities for optimizing the temperature dependences of PV devices performances. While some of them come naturally with improvements in V_{oc} for example, other approaches consist in optimizing the performances in specific operating conditions at the expense of the performances in the standard test conditions.

Appendix 1

Analogy Between a Photovoltaic Cell and a Leaky Bucket Feeding a Water Wheel

The analogy illustrated in Fig. 2.21 is inspired by a lecture from Prof. Yablonovitch at the International School of Solid State Physics in 2012. The version presented here is modified to explain better certain physical mechanisms of solar cell operation. In this analogy, a photovoltaic device is compared with a leaky

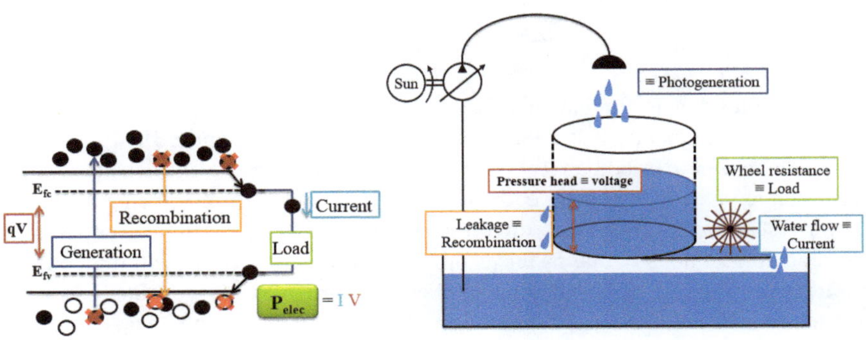

Fig. 2.21 Hydraulic analogy of the operating principles of a solar cell

bucket of water. This leaky bucket is connected to a water wheel whose mechanical resistance represents the electrical resistance of the load connected to the solar cell. The power dissipated in the wheel (≡output power) is equal to the product of the water flow, analog of the electrical current, and the pressure head in the bucket which compares to the voltage across the solar cell. Water flows into the bucket by a hose connected to a pump powered by the Sun. The water droplets represent the electrons excited into the conduction band by the incident solar photons. The bucket leaks through its sidewalls to represent the recombination of excited carriers. The more water is in the bucket, the more water leaks from the sides just like recombination rates increase when the cell voltage increases. This analogy illustrates that the amount of water in the bucket (≡cell voltage) depends on the balance between water flowing in by the hose (≡photogeneration) and water flowing out by its sides or through the water wheel (≡recombination and output current). To transfer the maximum power into the wheel, its resistance needs to be adjusted so that it is not too low to maintain some pressure in the bucket but not too high to prevent excessive leaking from the bucket. When the resistance of the waterwheel is negligible—short-circuit configuration—the bucket does not fill and most of the water entering through the shower head exits directly through the waterwheel ($J_{sc} \approx J_{photogenerated}$). When the outlet to the waterwheel is closed—open-circuit configuration—the bucket fills with water until the flow leaking through the sides equals that entering from the shower head. This explains why the open-circuit voltage indicates the balance between photogeneration and recombination rates.

Appendix 2

Temperature Dependence of the Diode Saturation Current

The following expression is often used for the diode saturation current density of a solar cell (Green 1982):

$$J_0 = C\,T^{\gamma} \exp\left(-\frac{E_{g0}}{k\,T_c}\right). \tag{2.56}$$

This expression is fundamental as it is a basis for many works dealing with temperature coefficients (Fan 1986; Nell and Barnett 1987; Landis 1994; Yoon and Garboushian 1994; Friedman 1996). The diode saturation current density, J_0, is a fundamental parameter of solar cells. It corresponds to the diffusion current density of carriers in the reverse diode direction when no voltage is applied, i.e. it is the diffusion current density that balances the drift current density at the p-n junction in equilibrium (Fig. 2.22a). Under an applied forward bias V, the potential at the junction is lowered and the concentration of carriers with enough energy to diffuse

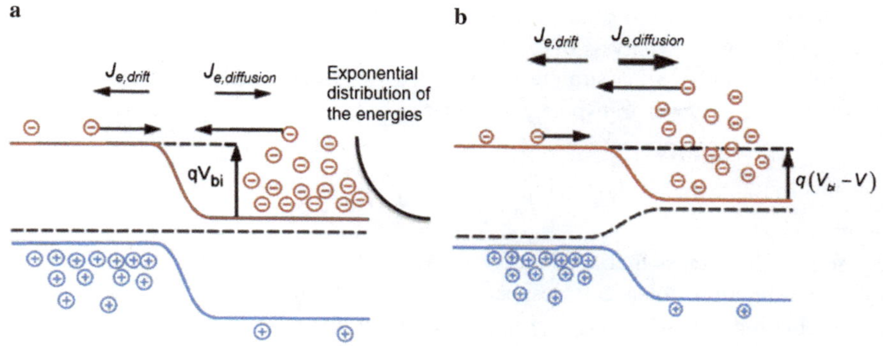

Fig. 2.22 p-n junction diagram in the dark under **a** no bias, **b** an external bias V. Figure adapted from a document by Bill Wilson under a Creative Commons Attribution License http://cnx.org/contents/8c8d95ff-8ac8-421c-a009-6bff6b862903@14/P-N_Junction:_Part_I

across it, is increased by $\exp(qV/kT_c)$ (Fig. 2.22b). This explains the form of the ideal diode law (Shockley 1948):

$$J = J_0\left(\exp\left(\frac{qV}{kT_c}\right) - 1\right). \tag{2.57}$$

The principle of superposition states that for a linear system, the response to several excitations is equal to the sum of the responses caused by each individual excitation. In the case of a solar cell, the excitations can be caused by light and/or voltage. Assuming linearity (which is usually valid under low injection conditions, i.e. when majority carrier concentrations are much larger that minority carrier concentrations), the principle of superposition states that the current density that can be extracted from a solar cell (J) equals that photogenerated (J_{ph}) minus that recombined which corresponds to the current density flowing in reverse bias through the diode in the dark:

$$J = J_{ph} - J_0\left(\exp\left(\frac{qV}{kT_c}\right) - 1\right) \tag{2.58}$$

This expression illustrates the name commonly given to the parameter J_O (diode or dark saturation current density). Indeed, in the dark and under a large reverse bias, the current saturate towards the value J_O. However, the parameter J_O has a more profound physical signification: it is proportional to the recombination rate in thermal equilibrium (with no applied bias). It is used to describe the strength of the recombination mechanisms independent of the bias condition. Because the value of the bias corresponds to the magnitude of the disequilibrium, the recombination rate can often be written as the product of J_O and $\exp(qV/kT_c)$. However, the situation can be more complex because different recombination mechanisms have different bias dependences. There is an interesting discussion on the physical meaning of the parameter J_O and an analysis of different recombination scenarios in Cuevas (2014).

Under a number of assumptions (detailed in Green 1982), J_O can be derived easily from the carrier transport equations:

$$J_0 = q\left(\frac{D_e}{L_e}\frac{1}{N_A}F_P + \frac{D_h}{L_h}\frac{1}{N_D}F_N\right) n_i^2 \tag{2.59}$$

F_P and F_N are factors that account for the finite dimensions of the P and N regions of a pn junction solar cell respectively and are thus function of the surface recombination velocities. $D_{e,h}$ and $L_{e,h}$ are the diffusion coefficients and the diffusion lengths of electrons and holes. $N_{A,D}$ are the doping concentrations of acceptors and donors. n_i is the intrinsic carrier concentration which is maybe the most crucial parameter in cell modeling. It corresponds to the concentration of carriers that are thermally excited from the valence band to the conduction band in an intrinsic, i.e. non-doped, material. For a non-degenerate semiconductor, i.e. where the Fermi energy is at least $3kT_c$ away from either band edge, the product of the electron and hole intrinsic concentrations is given by Green (1982):

$$n_i^2 = N_c N_v \exp(-E_g/kT_c) \tag{2.60}$$

where N_c and N_v are the number of effectively available states in the conduction and valence bands.

Comparing (2.56) with (2.59) and (2.60) shows that the term γ describes the temperature dependence of a large number of parameters (namely N_c, N_v, F_P, F_N, D_e, D_h, L_e, L_h, N_A and N_D).

The densities of states can be calculated by solving the Schrödinger equation assuming the simple particle-in-a-box model. Using the Boltzmann probability function (instead of the Fermi-Dirac probability function in the case of a non-degenerate semiconductor), the product of effectively available states in the conduction and valence bands, N_c and N_v, can be written as:

$$N_c N_v = 4\,(2\,\pi\,k\,T_c/h^2)^3 (m_e^* m_h^*)^{3/2} \tag{2.61}$$

where m_e* and m_h* are the effective masses of the electrons and the holes.

F_P and F_N are temperature dependent because of the temperature dependence of the surface recombination velocities.

The diffusion coefficients D_e, D_h are related to the carrier mobilities by the Einstein relations:

$$D_{e,h} = \mu_{e,h}\frac{k\,T_c}{q}. \tag{2.62}$$

The thermal energy kT_c which causes the random motion of carriers is logically present in the equation of the diffusion coefficient. Also, the mobilities of the carriers are function of temperature. These temperature dependences are function of the doping concentrations because doping impacts the carrier equilibrium concentrations (Klaassen 1992a, b).

The diffusion lengths (L_e, L_h) correspond to the average distances traveled by excited carriers before their recombination. They depend on the diffusion coefficients (D_e, D_h) and the lifetimes of the carriers (τ_e, τ_h):

$$L_{e,h} = \sqrt{D_{e,h}\tau_{e,h}}. \tag{2.63}$$

Lifetimes are also temperature dependent because the different existing recombination mechanisms are function of the cell temperature (Huldt et al. 1979; Schenk 1992; Trupke et al. 2003).

Eventually, the dopant concentrations, N_A and N_D, may depend on temperature in the case of incomplete ionization of the dopants in the range of temperature of interest.

The number of temperature dependent parameters listed above illustrates the complexity of the temperature dependence of the diode saturation current density. In several works (e.g. Nell and Barnett 1987; Friedman 1996), the temperature dependence of $\frac{D_e}{L_e}\frac{1}{N_A}F_P$ in (2.59) is neglected and it is assumed that the temperature dependence of J_O is driven only by that of n_i^2 (neglecting possible temperature dependences of the effective masses). This gives:

$$J_0 = C\,T_c^3 \exp\left(-\frac{E_{g0}}{k\,T_c}\right) \tag{2.64}$$

where C is a constant assumed independent of the temperature.

Additional parameters are sometimes used in the diode saturation current density equation (Fan 1986; Yoon and Garboushian 1994). For example, the following equation, initially proposed by Fan, is the basis for several works on temperature coefficients (Fan 1986):

$$J_0 = C\,T_c^{3/a} \exp\left(-\frac{E_g}{b\,k\,T_c}\right) \tag{2.65}$$

where C is an empirical parameter related to the cell material and a,b are "empirical parameters depending on the quality of the cell material and junction" (Fan 1986). These parameters are sometimes extracted from fits to experimental data and are not directly correlated to any physical mechanism. However, such analyses can lead to systematic errors in the modeling (Green 2003).

It is noteworthy that the expression of the diode saturation current given by (2.59) corresponds to a scenario where recombinations mainly occur in the quasi-neutral regions, i.e. in the bulk, or at the surfaces. There exist several expressions derived for different limiting scenarios. For example, if the recombinations are mostly happening in the depletion region, the diode saturation current can be written as (Siefer and Bett 2012):

$$J_{02} = \frac{n_i\, W_d\, k\, T_c}{2(V_d - V)\tau_{n0}} \tag{2.66}$$

where W_d is the width of the depletion zone, V_d is the diffusion voltage and τ_{n0} is the minimum electron lifetime. This expression shows a different temperature dependence from (2.59). A discussion of different possible scenarios is presented in Sect. 2.3.1.

References

Ali B, Murray R, Hegedus SS, Ismat Shah S (2012) Analysis of voltage and temperature dependent photocurrent collection in P3HT/PCBM solar cells. J Appl Phys 112:1–11. Doi:10.1063/1.4768910

Araújo GL, Martí A (1994) Absolute limiting efficiencies for photovoltaic energy conversion. Sol Energy Mater Sol Cells 33:213–240. Doi: 10.1016/0927-0248(94)90209-7

Baker-Finch SC, McIntosh KR (2010) A freeware program for precise optical analysis of the front surface of a solar cell. In: 35th IEEE photovoltaic specialists conference (PVSC). IEEE, pp 2184–2187

Baldasaro PF, Raynolds JE, Charache GW et al (2001) Thermodynamic analysis of thermophotovoltaic efficiency and power density tradeoffs. J Appl Phys 89:3319–3327. Doi:10.1063/1.1344580

Battaglia C, De Nicolas SM, De Wolf S et al (2014) Silicon heterojunction solar cell with passivated hole selective MoOx contact. Appl Phys Lett 104:2012–2017. Doi:10.1063/1.4868880

Beard MC (2011) Multiple exciton generation in semiconductor quantum dots. J Phys Chem Lett 2:1282–1288. Doi:10.1021/jz200166y

Boriskina SV, Green MA, Catchpole K et al (2016) Roadmap on optical energy conversion. J Opt 18:73004. Doi:10.1088/2040-8978/18/7/073004

Brandt RE, Mangan NM, Li JV et al (2016) Temperature- and intensity-dependent photovoltaic measurements to identify dominant recombination pathways. In: Proceedings of the 43rd IEEE photovoltaic specialist conference

Brown AS, Green MA (2002) Impurity photovoltaic effect: fundamental energy conversion efficiency limits. J Appl Phys 92:1329–1336. Doi:10.1063/1.1492016

Cuevas A (2014) The recombination parameter J_0. Energy Procedia 0:53–62. Doi:10.1016/j.egypro.2014.08.073

D'Innocenzo V, Grancini G, Alcocer MJP et al (2014) Excitons versus free charges in organo-lead tri-halide perovskites. Nat Commun 5:1–6. Doi:10.1038/ncomms4586

Deceglie MG, Silverman TJ, Marion B, Kurtz SR (2014) Metastable changes to the temperature coefficients of thin-film photovoltaic modules. In: 40th IEEE photovoltaic specialist conference. IEEE, Honolulu, HI, pp 337–340

Dey P, Paul J, Bylsma J et al (2013) Origin of the temperature dependence of the band gap of PbS and PbSe quantum dots. Solid State Commun 165:49–54. Doi:10.1016/j.ssc.2013.04.022

Dupré O (2015) Physics of the thermal behavior of photovoltaic devices. PhD thesis. INSA Lyon, France

Dupré O, Vaillon R, Green MA (2015a) Experimental assessment of temperature coefficient theories for silicon solar cells. IEEE J Photovolt 1–5. Doi:10.1109/JPHOTOV.2015.2489864

Dupré O, Vaillon R, Green MA (2015b) Physics of the temperature coefficients of solar cells. Sol Energy Mater Sol Cells 140:92–100. Doi:10.1016/j.solmat.2015.03.025

Emery K (2008) Photovoltaic test performance: laboratory test procedures measure photovoltaic cells and modules according to international standards. Photonics Spectra 42:76–80

Emery K, Burdick J, Caiyem Y et al (1996) Temperature dependence of photovoltaic cells, modules and systems. In: 25th IEEE photovoltaic specialists conference. IEEE, pp 1275–1278

Even J, Pedesseau L, Dupertuis MA et al (2012) Electronic model for self-assembled hybrid organic/perovskite semiconductors: reverse band edge electronic states ordering and spin-orbit coupling. Phys Rev B-Condens Matter Mater Phys 86:3–6. Doi:10.1103/PhysRevB.86.205301

Fan J (1986) Theoretical temperature dependence of solar cell parameters. Sol cells 17:309–315

Fraunhofer (2016) Photovoltaics report

Friedman DJ (1996) Modelling of tandem cell temperature coefficients. In: 25th IEEE photovoltaic specialists conference. IEEE, pp 89–92

Goni AR, Cantarero A, Syassen K, Cardona M (1990) Effect of pressure on the low-temperature exciton absorption in GaAs. Phys Rev B 41:10111–10119. Doi:10.1103/PhysRevB.41.10111

Green MA (1981) Solar cell fill factors: general graph and empirical expressions. Solid State Electron 2–3

Green MA (1982) Solar cells: operating principles, technology, and system applications. Prentice-Hall, Englewood Cliffs

Green MA (1990) Intrinsic concentration, effective densities of states, and effective mass in silicon. J Appl Phys 67:2944–2954. Doi:10.1063/1.345414

Green MA (2001) Third generation photovoltaics: ultra-high conversion efficiency at low cost. Prog Photovoltaics Res Appl 9:123–135. Doi:10.1002/pip.360

Green MA (2003) General temperature dependence of solar cell performance and implications for device modelling. Prog Photovoltaics Res Appl 11:333–340. Doi:10.1002/pip.496

Green MA (2006) Third generation photovoltaics. Springer, Berlin Heidelberg

Green MA (2012) Radiative efficiency of state of the art photovoltaic cells. Prog Photovolt Res Appl 20:472–476. Doi:10.1002/pip

Green MA, Emery K, Blakers AW (1982) Silicon solar cells with reduced temperature sensitivity. Electron Lett 18:97–98. Doi:10.1049/el:19820066

Green MA, Blakers AW, Osterwald CR (1985) Characterization of high-efficiency silicon solar cells. J Appl Phys 58:4402. Doi:10.1063/1.336286

Green MA, Ho-Baillie A, Snaith HJ (2014) The emergence of perovskite solar cells. Nat Photonics 8:506–514. Doi:10.1038/nphoton.2014.134

Green MA, Keevers MJ, Thomas I et al (2015) 40% efficient sunlight to electricity conversion. Prog Photovoltaics Res Appl 23:685–691. Doi:10.1002/pip.2612

Green MA, Emery K, Hishikawa Y, Warta W, Dunlop ED (2016) Solar cell efficiency tables (Version 48). Prog Photovoltaics Res Appl 24:905–913. Doi:10.1002/pip.2788

Grover S, Li JV (2015) Theory and analysis of temperature coefficient of open-circuit voltage (dV_{oc}/dT) in heterojunction solar cells. In: 42nd IEEE photovoltaic specialists conference, New Orleans

Grover S, Li JV, Young DL et al (2013) Reformulation of solar cell physics to facilitate experimental separation of recombination pathways. Appl Phys Lett. Doi:10.1063/1.4819728

Helmers H, Schachtner M, Bett AW (2013) Influence of temperature and irradiance on triple-junction solar subcells. Sol Energy Mater Sol Cells 116:144–152. Doi:10.1016/j.solmat.2013.03.039

Henry CH (1980) Limiting efficiencies of ideal single and multiple energy gap terrestrial solar cells. J Appl Phys 51:4494–4500. Doi:10.1063/1.328272

Hirst LC, Ekins-Daukes NJ (2011) Fundamental losses in solar cells. Prog Photovolt Res Appl 19:286–293. Doi:10.1002/pip.1024

Huldt L, Nilsson NG, Svantesson KG (1979) The temperature dependence of band-to-band Auger recombination in silicon. Appl Phys Lett 35:776. Doi:10.1063/1.90974

IEC (2008) Photovoltaic devices—part 3: measurement principles for terrestrial photovoltaic (PV) solar devices with reference spectral irradiance data, 2nd edn. In: Int. Stand., IEC 60904-3

Ishihara T (1994) Optical properties of PbI-based perovskite structures. J Lumin 60–61:269–274. Doi:10.1016/0022-2313(94)90145-7

Jiang Y, Soufiani AM, Gentle A et al (2016) Temperature dependent optical properties of $CH_3NH_3PbI_3$ perovskite by spectroscopic ellipsometry. Appl Phys Lett 108:61905. Doi:10.1063/1.4941710

Katz EA, Faiman D, Tuladhar SM et al (2001) Temperature dependence for the photovoltaic device parameters of polymer-fullerene solar cells under operating conditions. J Appl Phys 90:5343–5350. Doi:10.1063/1.1412270

Klaassen D (1992a) A unified mobility model for device simulation—I. Model equations and concentration dependence. Solid State Electron 35:953–959

Klaassen D (1992b) A unified mobility model for device simulation—II. Temperature dependence of carrier mobility and lifetime. Solid State Electron 35:961–967

Kosten EDE, Atwater JHJ, Parsons J et al (2013) Highly efficient GaAs solar cells by limiting light emission angle. Light Sci Appl 2:1–6. Doi:10.1038/lsa.2013.1

Kosten ED, Newman BK, Lloyd JV et al (2015) Limiting light escape angle in silicon photovoltaics: ideal and realistic cells. IEEE J Photovolt 5:61–69. Doi:10.1109/JPHOTOV.2014.2360566

Landis G (1994) Review of solar cell temperature coefficients for space. In: Proceedings of the 13th space photovoltaic research and technology conference, pp 385–399

Landsberg PT, Badescu V (2000) Carnot factor in solar cell efficiencies. J Phys D Appl Phys 33:3004–3008. Doi:10.1088/0022-3727/33/22/320

Li JV, Grover S, Contreras MA et al (2014a) A recombination analysis of Cu(In,Ga)Se2 solar cells with low and high Ga compositions. Sol Energy Mater Sol Cells 124:143–149. Doi:10.1016/j.solmat.2014.01.047

Li JV, Grover S, Repins IL et al (2014b) Electrical characterization of interface recombination and its dependence on band offset, potential barrier height, and inversion in certain heterojunction solar cells. In: 40th photovoltaic specialists conference. IEEE, pp 686–690. Doi:http://dx.doi.org/10.1109/PVSC.2014.6925013

Liou KN (2002) An introduction to atmospheric radiation. Academic Press, USA

Luque A, Martí A (1997) Increasing the efficiency of ideal solar cells by photon induced transitions at intermediate levels. Phys Rev Lett 78:5014–5017. Doi:10.1103/PhysRevLett.78.5014

Luque A, Martí A, Stanley C (2012) Understanding intermediate-band solar cells. Nat Photonics 6:146–152. Doi:10.1038/nphoton.2012.1

Markvart T (2008) Solar cell as a heat engine: energy-entropy analysis of photovoltaic conversion. Phys Status Solidi 205:2752–2756. Doi:10.1002/pssa.200880460

Martí A, Araújo GL (1996) Limiting efficiencies for photovoltaic energy conversion in multigap systems. Sol Energy Mater Sol Cells 43:203–222. Doi:10.1016/0927-0248(96)00015-3

Menard E, Meitl M, Burroughs S (2012) Indirect temperature measurement of CPV solar cells using wavelength shift of the sub-cells luminescence emission peaks. In: 27th European photovoltaic solar energy conference and exhibition, pp 189–193

Nell MME, Barnett AAM (1987) The spectral p-n junction model for tandem solar-cell design. IEEE Trans Electron Dev 34:257–266. Doi:10.1109/T-ED.1987.22916

Nozik AJ (2008) Multiple exciton generation in semiconductor quantum dots. Chem Phys Lett 457:3–11. Doi:10.1016/j.cplett.2008.03.094

Olkhovets A, Hsu RC, Lipovskii A, Wise F (1998) Size-dependent temperature variation of the energy gap in lead-salt quantum dots. Phys Rev Lett 81:3539–3542. Doi:10.1103/PhysRevLett.81.3539

Pässler R (1999) Parameter sets due to fittings of the temperature dependencies of fundamental bandgaps in semiconductors. Phys Status Solidi 216:975–1007. Doi:10.1002/(SICI)1521-3951(199912)216:2<975:AID-PSSB975>3.0.CO;2-N

Philipps SP, Hoheisel R, Gandy T et al (2011) An experimental and theoretical study on the temperature dependence of GaAs solar cells. In: 37th IEEE photovoltaic specialists conference, pp 001610–001614. Doi:10.1109/PVSC.2011.6186264

Ponce-Alcántara S, Connolly JP, Sánchez G et al (2014) A statistical analysis of the temperature coefficients of industrial silicon solar cells. Energy Procedia 55:578–588. Doi:10.1016/j.egypro.2014.08.029

Press WH (1976) Theoretical maximum for energy from direct and diffuse sunlight. Nature 264:734–735

Richter A, Hermle M, Glunz SW (2013) Reassessment of the limiting efficiency for crystalline silicon solar cells. IEEE J Photovoltaics 3:1184–1191. Doi:10.1109/JPHOTOV.2013.2270351

Riedel I, Parisi J, Dyakonov V et al (2004) Effect of temperature and illumination on the electrical characteristics of polymer-fullerene bulk-heterojunction solar cells. Adv Funct Mater 14:38–44. Doi:10.1002/adfm.200304399

Riesen Y, Stuckelberger M, Haug F-J et al (2016) Temperature dependence of hydrogenated amorphous silicon solar cell performances. J Appl Phys 119:44505. Doi:10.1063/1.4940392

Ross RT (1967) Some thermodynamics of photochemical systems. J Chem Phys 46:4590. Doi:10.1063/1.1840606

Ross RT, Nozik AJ (1982) Efficiency of hot-carrier solar energy converters. J Appl Phys 53:3813–3818. Doi:10.1063/1.331124

Ruppel W, Würfel P (1980) Upper limit for the conversion of solar energy. IEEE Trans Electron Devices 27:877–882. Doi:10.1109/T-ED.1980.19950

Schenk A (1992) A model for the field and temperature dependence of Shockley-Read-Hall lifetimes in silicon. Solid State Electron 35:1585–1596

Seif JP, Descoeudres A, Filipic M et al (2014) Amorphous silicon oxide window layers for high-efficiency silicon heterojunction solar cells. J Appl Phys. Doi:10.1063/1.4861404

Seif JP, Krishnamani G, Ballif C, De Wolf S (2015) Amorphous/ crystalline silicon interface passivation: ambient-temperature dependence and implications for solar cell performance. IEEE J Photovolt 5:718–724

Seif JP, Menda D, Descoeudres A et al (2016) Asymmetric band offsets in silicon heterojunction solar cells: impact on device performance. J Appl Phys 120:54501. Doi:10.1063/1.4959988

Shockley W (1948) The theory of p-n junctions in semiconductors and p-n junction transistors. Bell Syst Tech J 435–489. Doi:10.1002/j.1538-7305.1949.tb03645.x

Shockley W, Queisser HJ (1961) Detailed balance limit of efficiency of p-n junction solar cells. J Appl Phys 32:510–519. Doi:10.1063/1.1736034

Siefer G, Bett AW (2012) Analysis of temperature coefficients for III-V multi-junction concentrator cells. Prog Photovoltaics Res Appl 22:515–524. Doi:10.1002/pip

Siefer G, Abbott P, Baur C et al (2005) Determination of the temperature coefficients of various III–V solar cells. In: 20th European photovoltaic solar energy conference. Barcelona, Spain, pp 495–498

Singh P, Ravindra NM (2012) Temperature dependence of solar cell performance—an analysis. Sol Energy Mater Sol Cells 101:36–45. Doi:10.1016/j.solmat.2012.02.019

Singh P, Singh S, Lal M, Husain M (2008) Temperature dependence of I-V characteristics and performance parameters of silicon solar cell. Sol Energy Mater Sol Cells 92:1611–1616. Doi:10.1016/j.solmat.2008.07.010

Søndenå R, Berthod C, Odden JO et al (2015) Temperature dependent quantum efficiencies in multicrystalline silicon solar cells. Energy Procedia 77:639–645. Doi:10.1016/j.egypro.2015.07.093

Strevel N, Trippel L, Gloeckler M, Solar F (2012) First solar: greater energy yields in high-temperature conditions performance characterization and superior energy yield of first solar PV power plants in high-temperature conditions. Photovolt Int 1–6

Sze S, Ng K (1981) Physics of semiconductor devices. Wiley, New York

Taguchi M, Maruyama E, Tanaka M (2008) Temperature dependence of amorphous/crystalline silicon heterojunction solar cells. Jpn J Appl Phys 47:814–818. Doi:10.1143/JJAP.47.814

Tanay F, Dubois S, Enjalbert N, Veirman J (2011) Low temperature-coefficient for solar cells processed from solar-grade silicon purified by metallurgical route, pp 966–972. Doi:10.1002/pip

Tiedje T, Yablonovitch E, Cody GD, Brooks BG (1984) Limiting efficiency of silicon solar cells. IEEE Trans Electron Devices 31:711–716. Doi:10.1109/T-ED.1984.21594

Trupke T, Green MA, Würfel P (2002) Improving solar cell efficiencies by up-conversion of sub-band-gap light. J Appl Phys 92:4117–4122. Doi:10.1063/1.1505677

Trupke T, Green MA, Würfel P et al (2003) Temperature dependence of the radiative recombination coefficient of intrinsic crystalline silicon. J Appl Phys 94:4930–4937. Doi:10.1063/1.1610231

Varshni YP (1967) Temperature dependence of the energy gap in semiconductors. Physica 34:149–154. Doi:10.1016/0031-8914(67)90062-6

Virtuani A, Pavanello D, Friesen G (2010) Overview of temperature coefficients of different thin film photovoltaic technologies. In: 25th European photovoltaic solar energy conference

Vurgaftman I, Meyer JR, Ram-Mohan LR (2001) Band parameters for III-V compound semiconductors and their alloys. J Appl Phys 89:5815–5875. Doi:10.1063/1.1368156

Wu K, Bera A, Ma C et al (2014) Temperature-dependent excitonic photoluminescence of hybrid organometal halide perovskite films. Phys Chem Chem Phys 16:22476–22481. Doi:10.1039/C4CP03573A

Würfel P (1982) The chemical potential of radiation. J Phys C: Solid State Phys 15:3967–3985. Doi:10.1088/0022-3719/15/18/012

Würfel P (2009) Physics of solar cells: from basic principles to advanced concepts. Wiley, New York

Xiao C, Yu X, Yang D, Que D (2014) Impact of solar irradiance intensity and temperature on the performance of compensated crystalline silicon solar cells. Sol Energy Mater Sol Cells 128:427–434. Doi:10.1016/j.solmat.2014.06.018

Xu P, Chen S, Xiang H-J et al (2014) Influence of defects and synthesis conditions on the photovoltaic performance of perovskite semiconductor CsSnI3. Chem Mater 26:6068–6072. Doi:10.1021/cm503122j

Yoon S, Garboushian V (1994) Reduced temperature dependence of high-concentration photovoltaic solar cell open-circuit voltage (Voc) at high concentration levels. In: 1st world conference on photovoltaic energy conversion—WCPEC (A Joint Conference of PVSC, PVSEC and PSEC). IEEE, pp 1500–1504

Yu C, Chen Z, Wang J et al (2011) Temperature dependence of the band gap of perovskite semiconductor compound CsSnI3. J Appl Phys 110:63526. Doi:10.1063/1.3638699

Zhao J, Wang A, Robinson SJ, Green MA (1994) Reduced temperature coefficients for recent high performance silicon solar cells. Prog Photovoltaics Res Appl 2:221–225. Doi:10.1002/pip.4670020305

Chapter 3
A Thermal Model for the Design of Photovoltaic Devices

Abstract This chapter describes a thermal model for the design of photovoltaic devices, i.e. a method for calculating the heat generated within the device and that exchanged with the environment, from which the operating temperature can be determined, and subsequently the electrical power output at that temperature. Such a thermal model is necessary to go beyond standard test conditions (STC) in the design of PV devices. Indeed, two different cells having identical efficiencies at room temperature may perform quite differently in real operating conditions, because of dissimilarities in internal heat sources, cooling capabilities, and/or temperature coefficients. A set of analytical expressions is provided for each component of the heat source, with a brief introduction of mitigation options. The thermal balance of the PV device drives its equilibrium temperature. The dependence of that temperature on voltage and its consequences are highlighted. In particular, for solar cells, it is concluded that the normal operating condition temperature (NOCT) could be calculated at the maximum power point and be defined for different sets of typical climate and mounting conditions. This chapter also defines thermal design rules to optimize the performances of PV devices as a function of operating conditions. Certain options for minimizing the heat load, such as sub-bandgap and high-energy photon filtering, optimization of bandgap and other device parameters, are presented with selected examples.

3.1 Why a Thermal Model for Photovoltaic Devices?

A photovoltaic (PV) device converts power originating from an electromagnetic radiation source into electrical power. The remaining power is lost to the environment or transformed into heat within the device. Radiative losses to the environment are the reflected and transmitted radiation and the radiation emitted through radiative recombination within the device that eventually exits the cell.[1]

[1]This phenomenon is called external electroluminescence.

O. Dupré et al., *Thermal Behavior of Photovoltaic Devices*,
DOI 10.1007/978-3-319-49457-9_3

To date, there are many methods for limiting the reflection losses: anti-reflection coatings (Chen 2001), plasmonics (Atwater and Polman 2010; Pillai and Green 2010), etc. Radiative recombination is the process opposite to interband absorption and is therefore unavoidable in a PV cell. As paradoxical as it may seem, external electroluminescence should be promoted to its highest possible level to achieve maximum conversion efficiency (Green 2012; Miller et al. 2012). In general, the radiative losses are efficiently mitigated by optical means. Considering the actual efficiencies of solar cells, comprised between a few percent and up to 46%,[2] this means that a large fraction of the incident radiation is converted into heat. As a result, solar cells do not operate at ambient temperature, but most often several dozen degrees above (see Sect. 1.1). However, the efficiency of solar cells is currently rated in standard test conditions (STC) which correspond, inter alia, to a temperature of 25 °C. Thus, the energy yield of solar cells depends on their temperature coefficient and on their operating temperature. In Chap. 2, temperature coefficients have been discussed in detail: while engineering can improve them, they will remain non-negligible. Cell operating temperature is also a limiting factor, and thus must be known with accuracy for energy yield prediction and preferably be lowered at small energy cost (see Chap. 1). This temperature depends on the amount of thermal power generated within the device and on how much of it can be evacuated through thermal exchanges with a cooler environment. These heat powers generated and evacuated are governed by the design of the device. One question, not often raised, is about the dependence of the heat generated within the cell on its composition and structure (e.g. semiconductors used, layer thicknesses and doping, type of contacts, etc.). Two different PV modules might have the same efficiency at room temperature, but quite different ones in operation, because of unequal temperature coefficients (variations of the losses with temperature, see an illustration in Fig. 2.20), but also because the heat generated internally might be different, and their ability to evacuate that heat to the same environment different as well.

As a result, it is of uppermost interest to be able to evaluate the heat source generated within the PV device with exactitude, and to be able to link each component of this heat source with the device design parameters, for optimizing its performances for its operating conditions in the field rather than for the STC. Also, depending on the thermal operating conditions, in order to optimize the cooling it is essential to quantify how much heat can be evacuated as a function of the device design parameters. Thus the general thermal model presented in this chapter aims at establishing a method for calculating both the heat generated within the device and that exchanged with the environment, from which the temperature of the device can be determined, and subsequently the electrical output at that temperature.

[2]46% is for the best research solar cell on the NREL graph dated August 2016.

3.2 Model for the Heat Source

A detailed model for evaluating the internal heat source can be built by taking into consideration each conversion loss mechanism resulting in heat generation (Dupré et al. 2016). The resulting analytical expressions are summarized in Table 3.1 for single-junction solar cell under an illumination defined as a function of photon energy E by a photon flux density $PFD(E)$. The device operates at a current density J (A m^{-2}) and voltage V (V) on the J-V characteristic and at a temperature T_c.

Such a comprehensive thermal model can be interpreted from two viewpoints: that of the physics governing each component and correlatively that of engineering for device design toward achieving better performance. In the following, these viewpoints are illustrated for each component of the total heat source. First, it should be noticed that the first three terms (sub-bandgap, parasitic absorption and thermalization) do not depend on the operating point on the J-V characteristic. A convenient way to understand what these heat sources correspond to is to observe the dual axis representation given in Fig. 3.1. The dashed line represents the cumulated photon density of the AM1.5 solar spectrum times the elementary electrical charge[3] and the full line represents the radiation flux absorbed by the cell, i.e. neither reflected nor transmitted.[4] The optical losses represent a shortage for the conversion of photon energies above the bandgap. The "parasitic" heat source corresponds to the photons with energy larger than or equal to the bandgap energy (i.e. that could potentially contribute to charge carriers generation) that are absorbed by the free carriers and the atomic lattice. In the example of the crystalline silicon (c-Si) solar cell depicted in Fig. 3.1, it is negligible. However, photons with less energy than the bandgap cannot be absorbed in an interband process and are thus mostly absorbed through parasitic processes which contribute to heating the cell. In the example of a c-Si cell, the "sub-bandgap" heat source is approximately 15% of the incoming radiation power. The most obvious option for eliminating this heat source is to filter out these low energy photons so that they are not absorbed by the cell. Otherwise, a remedy is the up-conversion of these photons, which consists in combining two sub-bandgap photons into a photon with sufficient energy for generating an electron-hole pair (Trupke et al. 2002a). Another possible mitigation of the "sub-bandgap" heat source comes with the multi-junction concept (Jackson 1955; Martí and Araújo 1996). The idea is to use several cells with different bandgaps so as to better cover the incident radiation spectrum and in turn increase photo-conversion efficiency. The spectrum-splitting and photon selectivity can be operated by arranging the cells stacked on top of one another and connected in series or in parallel (Green 2003), or via parallel illuminations of the cells on selected spectral ranges by optical means (Barnett et al. 2009; Green and Ho-Baillie 2010). A last example of possible

[3] At the bandgap energy, this gives the short-circuit current density assuming that every incident photon is converted into an electron-hole pair which gets collected (EQE = 1).

[4] At the bandgap energy, this corresponds to the short-circuit current density assuming that every absorbed photon is converted into an electron-hole pair which gets collected (IQE = 1).

Table 3.1 Losses generating heat, examples of mitigation solutions of these losses and analytical expressions of the corresponding heat sources (Q, in W m^{-2}). The cell is operating at the ($J(V)$, V) point on the J-V characteristic and at temperature T_c. It is supposed that the incident radiation can be modeled by a photon energy distribution function $PFD(E)$. $R(E)$ and $T(E)$ are respectively the reflectance and transmittance of the cell for photon energy E. f_{par}, f_{ib}, f_{ex} are the fraction of photons absorbed through parasitic processes, interband processes and emitted by radiative recombination that exit the cell, respectively.

Conversion loss coming with heat generation	Examples of mitigation solutions
Analytic expression of the corresponding heat source	
Sub-bandgap	Up-conversion and multi-junction devices, lowering absorption for $E < E_g$
$Q_{below\ E_g} = \int_0^{E_g} (1 - R(E) - T(E))\ PFD(E)\ E\,dE$	
Parasitic absorption of incident radiation	Lowering absorption by parasitic processes for $E > E_g$
$Q_{parasitic\ abs.} = \int_{E_g}^{\infty} (1 - R(E) - T(E))\, f_{par}(E)\ PFD(E)\ E\,dE$	
Thermalization	Down-conversion, hot-carrier, and multi-junction devices
$Q_{thermalization} = \int_{E_g}^{\infty} (1 - R(E) - T(E))\, f_{ib}(E)\ PFD(E)\ (E - E_g)\,dE$	
Parasitic absorption of photons emitted by radiative recombination	Lowering absorption by parasitic processes for $E > E_g$
$Q_{RR-parasitic\ abs.}(V) = \left(f_{par}/f_{ex}\right) P_{emission}(V)$ with $P_{emission}(V) = E_g \frac{2\,\Omega_{emit}}{c^2 h^3} \int_{Eg}^{\infty} \frac{\overline{EQE}\,E^2}{\exp\left(\frac{E-qV}{kT_c}\right)-1} dE$	
Non-radiative recombination of photons emitted by radiative recombination	Minimizing impurity and Auger recombinations
$Q_{NRR-RR}(V) = \left(f_{NRR}/f_{ex}\right) P_{emission}(V)$	
Current drop due to non-radiative recombinations	Minimizing impurity and Auger recombinations
$Q_{NRR-\Delta J}(V) = E_g\left(\frac{1}{ERE} - 1\right) \frac{2\,\Omega_{emit}}{c^2 h^3} \int_{Eg}^{\infty} \frac{\overline{EQE}\,E^2}{\exp\left(\frac{E-qV}{kT_c}\right)-1} dE$	
Current drop due to shunt losses	Maximizing shunt resistance
$Q_{shunts}(V) = \frac{V + J(V)\,R_s}{R_{sh}} \frac{E_g}{q}$	
Voltage drops	
$Q_{\Delta V}(V) = J(V)\left(\frac{E_g}{q} - V\right)$	
Carnot	
* $Q_{Carnot-bb}(V_{MPP}) = J_{MPP} \frac{E_g}{q} \frac{T_c}{T_s}$	
Angle mismatch	Matching cell radiation absorption-emission channels

(continued)

Table 3.1 (continued)

Conversion loss coming with heat generation		Examples of mitigation solutions
*	$Q_{am-bb}(V_{MPP}) = J_{MPP}\frac{kT_c}{q}\ln\left(\frac{\Omega_{emit}}{\Omega_{abs}}\right)$	
	Non-radiative recombination	Minimizing impurity and Auger recombinations
*	$Q_{NRR-\Delta V-bb}(V_{MPP}) = J_{MPP}\frac{kT_c}{q}\ln\left(\frac{1}{ERE_{MPP}}\right)$	
	Transport	Minimizing series resistance and minimizing currents
	$Q_{transport}(V) = R_s J^2(V)$	

E_g is the bandgap (in eV) of the cell material. $\overline{EQE}$ is the external quantum efficiency of the cell appropriately weighted over all angles of incident light and *ERE* is the External Radiative Efficiency, corresponding to the fraction of recombination events that results in the emission of a photon from the device (Green 2012). R_{sh} and R_s are the shunt and series resistance. *c*, *q*, *k* and *h* are respectively the light speed in vacuum, the elementary electrical charge, Boltzmann and Planck constants, and Ω_{emit} is the integral over the emission solid angle of the polar angle cosine defined as in Harder and Würfel (2003) and in Sect. 2.2. Expressions of the heat sources associated with voltage drops for Carnot, angle mismatch and non-radiative recombination losses (*), are given at the maximum power point (J_{MPP},V_{MPP}) when the source and the cell are assumed to be blackbody radiators. The total heat source generated is the sum of all these terms. Details available in Dupré et al. (2016a)

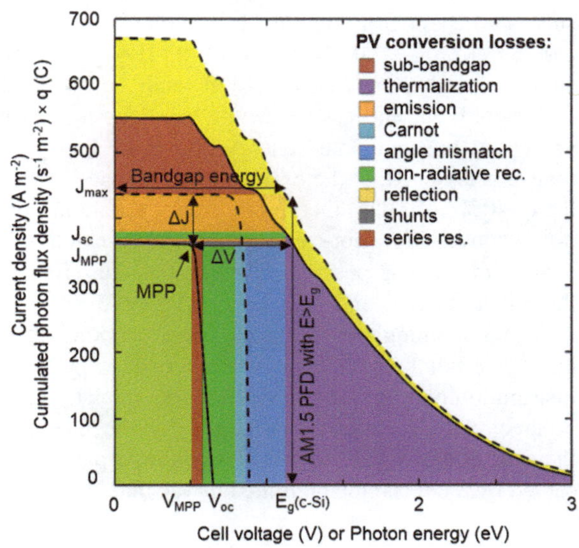

Fig. 3.1 AM1.5 cumulated photon flux density times electrical elementary charge [maximum potential current from: the incident photons (*dashed line*), the absorbed photons (*full line*)] as a function of photon energy; current density as a function of voltage (i.e. J-V characteristics). The case of a c-Si solar cell with a planar surface without anti-reflection coating is considered. Figure adapted from Dupré (2015)

sub-bandgap heat load reduction is the intermediate-band cell concept (Luque et al. 2012), where some photons of sub-bandgap energy are enabled to be absorbed and contribute to the photocurrent instead of generating heat. In any case, the remaining heat generation is governed by the cell material properties. For example, parasitic absorption by the free carriers depends on the semiconductor material doping. Heat dissipation is also driven by certain cell properties. For example, it is beneficial to have devices with a large emittance for the photon energy range where thermal emission occurs, i.e. in the infrared.[5] This cooling effect is included in the modeling of the flux exchanged by the device with the environment (see Sect. 3.3).

The photons that have an energy above the bandgap contribute more energy than required for generating an electron-hole pair. The difference to bandgap[6] is extra kinetic energy for the photo-generated electron and holes, which are named “hot carriers”. The excited carriers relax very quickly to the band edges, in about 10^{-12} s (Würfel 2009), through a process called “thermalization”. The excess energy is transferred from the carriers to the phonons through collisions with the lattice atoms. In the example involving a c-Si cell, the thermalization heat source represents nearly 30% of the incoming radiation power and thus is by far the largest. The corresponding thermal power depends on the photon energy distribution above the bandgap, which is function of the radiation source. One way to reduce this thermalization heat source is the down-conversion of the incident photons which consists in splitting the high-energy photons into more photons having just above the required energy for interband absorption (Trupke et al. 2002b). This enables to generate more than one electron-hole pair with a single high-energy photon. The concept of hot-carrier cells provides another solution for mitigating thermalization losses. The principle is to extract the hot carriers before they fully thermalize (Ross and Nozik 1982; Würfel 1997; Green 2003). It requires a fine understanding of the successive steps from photon absorption to the final thermalization of the generated carriers (König et al. 2010). Multi-junction cells provide a solution for reducing simultaneously the “thermalization” and “below bandgap” heat sources. The effect is two-fold: current loss below the band gap is reduced and with multiple bandgap absorbers, the cumulated energy above the bandgap that carriers eventually thermalize is lessened. In the example, the voltage independent heat sources reach more than 40% of the incoming solar radiation power. While up-conversion, down-conversion, hot-carrier, and multi-junction device concepts are under development, it is possible to come up with simple strategies to reduce some of these heat sources and their impacts (Sect. 3.5).

The remaining heat sources depend upon the applied voltage, thus on the operating point on the J-V characteristic. Fig 3.1 is helpful for visualizing the distribution and importance of each loss depicted as areas on the current density—voltage plot. The power harvested by the electrical carriers, assuming that all incident photons generate an electron-hole pair that does not recombine, is represented by the rectangle delimited by the maximum voltage (V_{max}) and the maximum

[5]The blackbody radiation spectrum peaks at around 10 μm at 300 K, while 99% of the incident radiation power are for wavelengths below 2.5 μm for the AM1.5 spectrum (see Figs. 2.10 and 3.6).

[6]When the carrier mean kinetic energy 3/2 kT can be neglected.

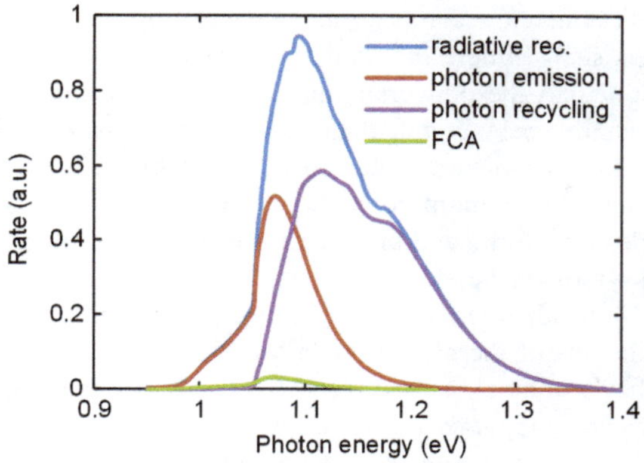

Fig. 3.2 Example of spectral rate of internal photon emission (radiative recombination or spontaneous emission or internal electroluminescence) and its decomposition into photon recycling (re-absorption through band-to-band transitions), free-carrier absorption (FCA), and external photon emission (external electroluminescence). Figure created with data from Richter et al. (2013)

current density (J_{max}).[7] From this unreachable[8] limit, each additional loss at the operating point ($J(V)$,V) can be conveniently expressed considering current density $\Delta J(V)$ and voltage $\Delta V(V)$ drops with the associated power losses $\Delta J(V)\ V_{max}$ and $J(V)\ \Delta V(V)$, respectively. The optical losses imply a current density drop without any heat generation except for parasitic absorption. Photons emitted within the cell by radiative recombination and that are not recycled, eventually escape the cell and cause a current density drop (Fig. 3.1). However, a fraction of the photons emitted within the cell is re-absorbed through parasitic processes by the free carriers and the atomic lattice or generate an electron-hole pair that recombine through non-radiative processes and thus contribute to heating the device (Fig. 3.1). In the example of Fig. 3.1, these two heat sources are negligible. Unlike radiative recombination, Auger and impurity recombination result directly in heat generation. The concept of external radiative efficiency [ERE, (Green 2012)] is convenient for formulating the heat source associated with the current drop due to non-radiative recombinations. As in Sect. 2.3, it involves the inverse of the ERE times the radiative recombination emission power ($P_{emission}$), without forgetting to remove the emission power itself (the −1 term) since it does not contribute to heating. The shunt losses come with a current drop and a corresponding heat source that is not exponentially but linearly dependent on voltage.

[7]At moderate doping and injection levels, V_{max} is limited by the bandgap (E_g/q, with E_g in J) (Green 1984) and J_{max} is given by the cumulated photon flux density for photon energy above the bandgap, times the elementary charge.

[8]Thermodynamics sets the Carnot limit which translates into a current (emission by radiative recombination) and a voltage drop, see Fig. 3.1.

As for the heat sources accompanying voltage drops, it is worth referring to fundamental limits first. Simple thermodynamic arguments involving energy and flux balances applied to the converter indicate that the Carnot limit[9] bounds its efficiency. From detailed balance calculations assuming that the radiation source and the device are blackbody radiators, it can be shown that radiative emission results in a drop of the voltage at maximum power point (Hirst and Ekins-daukes 2011). The expression includes the aforementioned Carnot factor and another term accounting for a possible dissymmetry between the source-to-the-cell and cell emission radiation channels.[10] Non-radiative (Auger and impurity) recombinations and transport losses due to collisions of the carriers with the lattice, are the cause of additional voltage drops. They can be expressed respectively using once again the concept of external radiative efficiency, and the Joule power law that involves the cell series resistance. As a result of these effects comes a cumulated voltage drop $\Delta V(V)$ which is simply $V_{max} - V$ and the associated heat source is $J(V)$ $(V_{max} - V)$.[11] These voltage drop heat sources can be minimized by reciprocating the radiation channels between the radiation source and the cell[12] (angle mismatch loss), minimizing Auger and impurity recombination (non-radiative recombination loss), and minimizing series resistance and avoiding too large currents (transport loss).

The physical mechanism for certain of the heat sources coming with voltage losses is known as Peltier heating. It happens when a current crosses an interface separating regions with different Peltier coefficients that are related to the average energy transported by the electrical carriers (Kittel 1996). In photovoltaic devices, this occurs at the selective membranes constituted by the junction and the peripheral contact regions, where sharp variations of band edges take place (Pipe et al. 2002; Breitenstein and Rakotoniaina 2005). Depending on current flow direction with respect to the electrical carrier energy gradient, this results locally in either cooling or heating. These thermoelectric effects impact the performances of light emitting diodes (LED) as well (Heikkilä et al. 2009; Xue et al. 2015).

The full expression of the heat power dissipated in a photovoltaic device contains eight components of which five depend on the applied bias. Adding these

---

[9] In the case of solar cells, assuming that the Sun is at a temperature T_s, this efficiency is $1 - T_c/T_s$. Assuming that the Sun is a blackbody radiator at 6000 K and the cell operates at 300 K, this maximum efficiency is 95%.

[10] This factor involves the natural logarithm of the ratio $\Omega_{emit}/\Omega_{abs}$, where Ω_{emit} is the projected solid angle in which the cell emits radiation and Ω_{abs} that under which the Sun is seen from the cell, see Sect. 2.1 or (Dupré et al. 2015). In the case of solar cells under one Sun illumination, where Ω_{emit} (π) is much larger than Ω_{abs} (6.87×10^{-5}), this induces the so-called "angle mismatch loss". This loss can be mitigated either by increasing Ω_{abs} via concentration (Hirst and Ekins-daukes 2011), or by reducing Ω_{emit} (Martí et al. 1997; Kosten et al. 2013) both by optical means. When the two radiation channels are reciprocal, this voltage drop vanishes.

[11] As $V_{max} = E_g/q$, this drop is also called bandgap-voltage offset (King et al. 2005, 2011).

[12] Formulated otherwise, for solar cells, it means that the solid angle under which the cell receives radiation from the source must be the same as the solid angle of emission of the cell. For thermophotovoltaic cells with near-field interactions (see Sect. 4.8), evanescent modes of radiation transfer have to be accounted for.

losses to the non-heat dissipating ones, the modeling of the heat sources described previously is found to be consistent by checking the overall power balance in the case of a typical c-Si cell (Dupré et al. 2016a).

3.3 Model for the Equilibrium Temperature

The common approaches for predicting the equilibrium temperature have been introduced in Sect. 1.2. This equilibrium is ruled by the thermal balance of the device. Here for the sake of generality, in steady state conditions, the expression given in Table 3.2 simply states that heat generated internally is balanced by the heat flux exchanged with the external media, via conduction, convection and radiation channels. It is also convenient to use a single heat transfer coefficient for conduction and convection. Specific configurations might require more elaborated heat diffusion and fluid mechanics modeling. For the sake of simplicity, the net radiation heat flux expression involves the spectral emittance of the system and blackbody spectral distributions, but once again the use of radiative transfer or electromagnetic modeling might be necessary in certain cases. For the thermal modeling of photovoltaic modules, layers encapsulating the photovoltaic cell and their thermal properties need to be taken into consideration, in particular the temperature variations across the module, with either opaque (Silva et al. 2010) or more elaborate semi-transparent radiation models (Holst et al. 2013; Vogt et al. 2015). Similarly, the performances of either a specifically designed or natural convective cooling system might depend on an ensemble of parameters that involves the converter constituents and its surrounding properties. Basics for a full modeling of heat transfer, out of the scope of the present book, can be found in Bergman et al. (2011) for general heat and mass transfer, and in Modest (2013) and Howell et al. (2015) for thermal radiation. As certain photovoltaic devices are constituted of particularly thin layers, it might in some cases be unavoidable to account for microscale and nanoscale thermal

Table 3.2 Steady-state heat equation that rules the equilibrium temperature (T_c) of a photovoltaic device. $Q(V)$ is the heat source generated in the device, h is a global heat transfer coefficient representing conductive and/or convective heat fluxes ($Q_{cond+conv}$) exchanged with the media surrounding the device (front and back surfaces) supposed to be at ambient temperature (T_{amb}). The radiation net flux (Q_{rad}) is driven by the spectral emittance ($\varepsilon(E)$) of the device, its temperature (T_c) and the temperature of the environment (T_e) that might need to be split into several components (e.g. for solar cells: the cold outer space, the clouds and the closer ground, nearby buildings or structures)

$Q(V) = Q_{cond+conv} + Q_{rad}$	
Conduction and convection	$Q_{cond+conv} = h(T_c - T_{amb})$
Radiation	$Q_{rad} = \int_0^{\infty} \varepsilon(E)\, E\, \left(PFD_{bb(T_c)}(E) - PFD_{bb(T_e)}(E)\right) dE$

Different ambient temperatures may need to be considered at the front and back surfaces (e.g. in the case of building integrated photovoltaics)

transport phenomena (Chen 2005; Pop 2010). Indeed, for semiconductor layers thinner than hundreds of nanometers, the thermal conductivity can be smaller than the bulk value owing to both ballistic transport and interface effects. Thermal resistances may also play a significant role at the interface between several elements of sandwiched layers in PV devices. This may imply at some point revisiting the macro-scale modeling described in the present book and replace it with thermal laws appropriate at the micro- and nano-scales (Volz 2007; Zhang 2007). As for transient analyses, the variations with time of internal energy has to be inserted in the expression of the thermal balance equation. Temporal changes of the operating conditions are incorporated in the derivation of the internal heat sources and heat flux exchanged. In the case of terrestrial photovoltaic systems, these are variations of illumination, wind speed, ambient and environment temperatures [see for example (Jones and Underwood 2001; Silva et al. 2010)].

3.4 Dependence on Voltage of the Heat Source and the Cell Temperature

3.4.1 Temperature Dependent Power Output as a Function of Voltage

The dependence on the applied voltage of the heat source $(Q(V))$ must be carefully examined. One obvious consequence is that the device equilibrium temperature depends on the voltage as well $(T_c(V))$. The dependence on voltage of the electrical output power $(P(V))$ should then be expressed as $P(V, T_c(Q(V, T_c(Q(\ldots)))))$. This illustrates clearly the previously mentioned coupling between the optical/radiative, electrical and thermal behaviors of the cell. This observation seems simple, yet it has some important consequences. In particular, the temperature of a device in given operating conditions will be different if the cell is open-circuited or working at its maximum power point. In other words, in real operating conditions, where the thermal and radiative conditions are set, the variations with voltage of the electrical output depend on temperature. The reason is that the heat source, which rules the device temperature, depends on the electrical power extracted from the device. This is illustrated in Fig. 3.3 where a calculation of the output electrical power is reported as a function of voltage taking into account that the temperature of the cell is a function of voltage. In this simulation, a state-of-the-art crystalline silicon solar cell with an external radiative efficiency of 0.57% is considered (Green 2012). The prescribed operating conditions are an AM1.5 illumination, a dissipation of the internally generated heat by convection only, with ambient air at 25 °C and a heat transfer coefficient equal to 20 W m^{-2} K^{-1}. For each voltage, the heat source is determined using the model described in Sect. 3.2. The equilibrium temperature at that voltage is calculated iteratively given that the internal heat source and the equilibrium temperature are interdependent. The current density $J(V)$ and power density $P(V)$ characteristics points are determined at the converged temperature

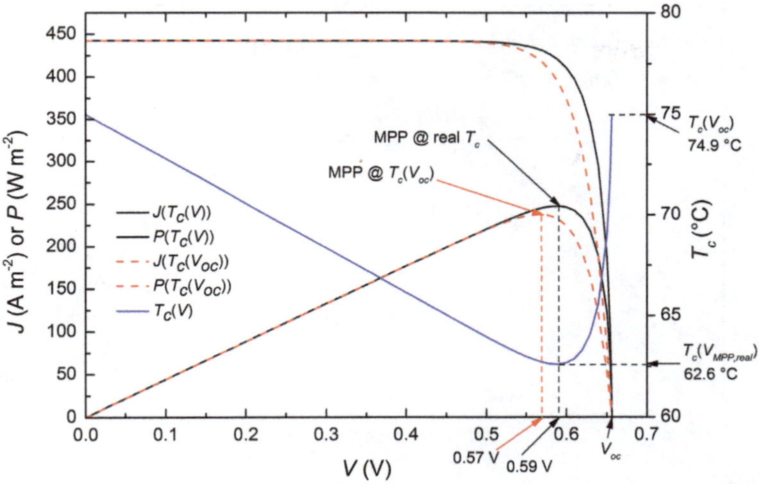

Fig. 3.3 Variations with voltage of the cell equilibrium temperature. J-V and P-V characteristics at real $T_c(V)$ and at fixed $T_c(V_{oc})$, for a crystalline silicon solar cell in a given operating condition (AM1.5 illumination, $T_{\mathrm{amb}} = 25$ °C, $h = 20$ W m^{-2} K^{-1}, $Q_{rad} = 0$ W m^{-2})

characteristic values $T_{\mathrm{c}}(V)$. This implicitly means that the superposition principle[13] must be applied separately for each voltage operating point.

Three contributions to the heat source do not depend on voltage: sub-bandgap absorption, parasitic absorption and thermalization (see Sect. 3.2). However, since the other heat sources depend on voltage, in given operating conditions, the heat source is larger at $V = V_{sc} = 0$ (short circuit) and $V = V_{oc}$ (open circuit) than at $V = V_{MPP}$. As a consequence, the same applies to the cell temperature. In short-circuit configuration, most of the photogenerated carriers do not recombine, but no electrical power can be extracted. Heat generation is dominated by Peltier and transport terms. Emission resulting from radiative recombination is usually negligible, in particular at zero voltage (see Table 3.1). In open-circuit configuration, photogeneration is counterbalanced by recombination and both electrical current and power are null. The non-radiative recombination processes contribute significantly to the heat source.[14] It is at the maximum power point that the heat source is minimum. This means that the temperature of the cell is smaller at the

[13]Usually, the superposition principle implies a separate calculation of the photocurrent (assumed to be independent of voltage) and the dark current as a function of voltage for a single temperature. Since the cell temperature depends on voltage, both the photocurrent and the dark current have to be recalculated at every voltage.

[14]Note that as the cells approach the radiative limit (*ERE* = 1), the fraction of recombinations that ends up in radiative emission becomes non negligible. Since the photons emitted by the cell do not contribute to heating the cell, this means that the heat source in open-circuit will diminish as the ERE increases towards unity and the temperature will eventually be lower in open-circuit than in short-circuit conditions.

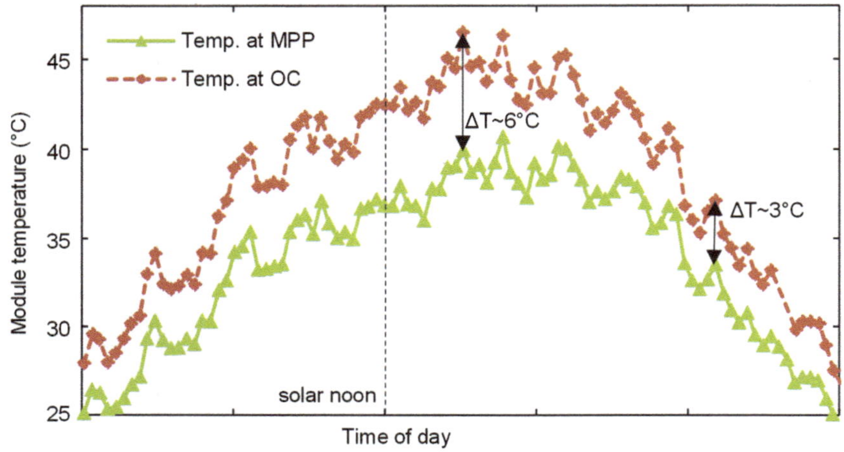

Fig. 3.4 Experimental observation of the difference in module temperature when the cell is operating either at open circuit (OC) or at the maximum power point (MPP). Figure created with data from Silva et al. (2010)

MPP than at open circuit (by 12.3 °C in the example of Fig. 3.3). This difference is experimentally confirmed in Silva et al. (2010) for PV modules operating in the field (see Fig. 3.4), however its magnitude is smaller ($\sim$6–7 °C at maximum). The different magnitudes may result from different operating conditions (incident power, heat transfer coefficients, etc.) but are also probably due to the different efficiencies of the devices considered. Indeed, the incident power that is converted in electricity corresponds to a reduction of the heat generated within the device.

On Fig. 3.3 are also plotted J-V and P-V characteristics calculated for a fixed cell temperature, equal to that determined in the chosen operating conditions at open circuit $(T_c(V_{oc}))$. Since the temperature in open circuit is larger than that at the MPP, it is found that applying the superposition principle for all voltages using a single temperature determined in open-circuit conditions, leads to J-V and P-V characteristics with a MPP slightly shifted to a lower value. By doing so, the output power at MPP is underestimated compared to that at the cell temperature in real operating conditions (the absolute difference in efficiency is 0.5%). In Silva et al. (2010), the absolute difference in efficiency is estimated to be comprised between 0.65 to almost 2%, depending on technology. The different values can be explained by differences in the temperature coefficients of the devices.

3.4.2 Revisiting the Definition of Nominal Operating Cell Temperature

The nominal operating cell temperature (NOCT) is defined for photovoltaic modules to provide an estimate of their performances in operating conditions different from the standard test conditions (STC), where in particular the cell temperature is prescribed (at 25 °C). The NOCT is the temperature reached by the open-circuited cells in a module mounted on an open-rack under a 800 W m^{-2} illumination with a wind velocity of 1 m s^{-1} and an ambient air temperature of 20 °C (Ross and Gonzalez 1980). These conditions were chosen so that the annual energy produced by a module is well approximated by its efficiency at NOCT, times the average irradiance incident on the module (in kWh/year) at the location of interest (Ross and Smokler 1986). However, it is worth noticing that the locations used in Ross and Gonzalez (1980) to define these conditions (Albuquerque, Cap. Hat., Omaha, Miami) are not representative of all the climates/operating conditions that can be found worldwide. Thus, in general, the NOCT is simply an indicator of the quality of the thermal design of a module, i.e. how efficiently it evacuates the heat generated internally.

From the observation made in Sect. 3.4.1, that a PV cell temperature is smaller by several degrees at MPP than in open circuit, it would be better to define the NOCT as the temperature reached by a module at its maximum power point rather than in open circuit (still in a given set of outdoor conditions) (Koehl et al. 2011; Dupré 2015). This would include the impact of the device conversion efficiency in the value of the NOCT. This alternative definition would correspond more closely to the temperature reached by the module in operation (and not in open circuit).

Additionally, since thermal (convective, conductive and radiative) boundary conditions and heat dissipation ability of modules are highly dependent on location and on the mounting configuration, it could be useful to define a set of NOCTs for several representative operating conditions. This would allow finer estimates of the annual energy productions.

3.5 Beyond Standard Test Conditions: Taking into Account Field Operating Conditions in the Design of Photovoltaic Devices

The thermal model described previously can be used to determine the heat generated within the device (Sect. 3.2), the heat flux that is exchanged with the surroundings and the equilibrium temperature that results from the balance between these two quantities (Sect. 3.3). As can be observed on the schematic of Fig. 3.5, the equilibrium temperature is governed by the operating conditions and by the device design. The operating conditions are the radiative and thermal boundary conditions of the modeled thermal system. The design of the device consists in its composition, its dimensions and the arrangement of its different parts. Optical/radiative, electrical and thermal properties stem from this design. Both operating conditions and device design rule the heat source (Q) and the heat fluxes exchanged (Q_{rad} and $Q_{conv\ +\ cond}$), thus the equilibrium temperature (T_c) and finally the maximum electrical power output (P_{MPP}) at that temperature via the temperature

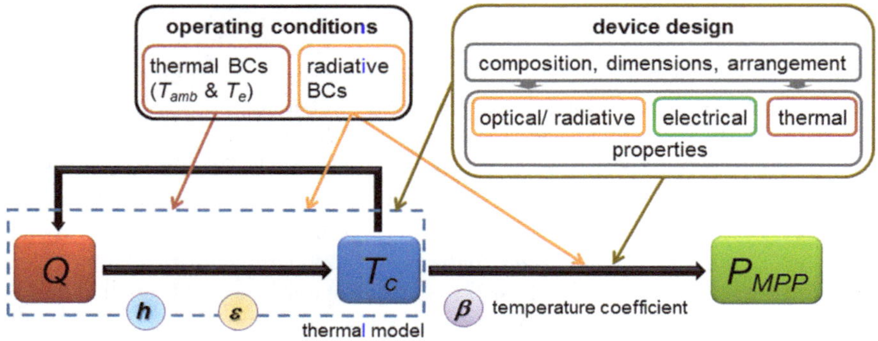

Fig. 3.5 Schematic of the thermal behavior of photovoltaic devices and its dependence on operating conditions and device design. Figure adapted from Dupré and Vaillon (2014)

coefficient (β). Intricate couplings may arise from dependences on temperature of optical/radiative, electrical and thermal properties. There is also a strong connection between the temperature coefficient—representing the variations with temperature of all losses—and the dependence of the heat source on temperature because most of the losses contribute to the heat source.

Nowadays, the efficiency of solar cells is rated in the so-called standard test conditions (STC), at a prescribed temperature of 25 °C. Accordingly, cells are primarily designed to perform best in these conditions. However, actual operating conditions significantly diverge from the STC (see Sect. 1.1), depending on the local climate and the mounting configuration. In Chap. 2, the temperature dependence of the electrical power output at the maximum power point has been discussed in details. This temperature driven performance of photovoltaic devices is depicted on the right side of the schematics of Fig. 3.5 via the so-called temperature coefficient. A thorough analysis of the physics ruling temperature coefficients (Sects. 2.2, 2.3 and 2.4) has shown that there are opportunities for designing solar cells that mitigate the loss of performance when temperature rises. The operating conditions may play a role on this design, for example through the impact of the spectral distribution of the illumination on the temperature coefficients (see Sects. 2. 2.1 and 2.3.2). At this stage, a design beyond STC means that two cells having the exact same performances in STC, might have quite diverging efficiencies at larger temperatures.

In addition, in the exact same operating conditions, these two cells may reach different temperatures. The nominal operating cell temperature (NOCT) somewhat tells about it. It is also an indicator of how much heat is generated internally. However, even if this nominal temperature, calculated for a very specific condition, can be extrapolated to other conditions (e.g. illuminations), it does not tell much on the opportunities that are available for minimizing the device temperature under actual operating conditions. From there, there are two levers to act on. The first one is to design a system that dissipates better the internally generated heat towards the environment. The second one is to reduce the heat generated internally. In general, a better conversion efficiency is synonymous with less losses and thus less heat sources. However, if this makes perfect sense as a general rule, there is no guarantee that a design leading to lower conversion losses in STC will be optimum in actual operating conditions. A detailed knowledge of each contribution to the heat source and its dependence on cell design and operating condition parameters is thus required.

In the following, the idea that the design of photovoltaic devices would benefit from taking actual operating conditions into consideration through a full thermal modeling (i.e. detailed calculation of the internal heat source and determination of the equilibrium temperature, as a function of operating conditions and device characteristics) is explored. In this approach, optimizing the performances of photovoltaic devices means adopting the following thermal design rules:

- *minimizing the internal heat load,*
- *maximizing the cooling of the device,*
- *minimizing the temperature coefficient,*

as a function of the operating conditions.

The resulting cell optimum characteristics are most likely to be different from those in STC, but also different from one operating condition to another (e.g. climate condition, mounting configuration, terrestrial vs space operation, etc.). The existing strategies to maximize heat dissipation toward the environment as a function of operating conditions have been presented in Sect. 1.3: convective cooling, hybrid photovoltaic/thermal solutions, radiative cooling, building integrated and floating solutions, etc. Possible solutions to lower the temperature coefficient have been examined in Sect. 2.4. The next sections present different designs of PV devices that aim at minimizing the internal heat load.

3.5.1 Sub-bandgap Energy Photon Filtering

In single-junction cells, the sub-bandgap energy photons are useless to the photo-conversion. The most obvious minimization of the sub-bandgap heat source is to prevent the cell from absorbing these photons.

One solution is to coat the cell with a layer that fully transmits the useful photons but reflects as much as possible the useless ones. An example of such filtering is given for GaAs solar cells used in space applications in Beauchamp et al. (1993), Beauchamp and Tuttle-Hart (1995). The near-infrared (NIR) radiation with wavelengths approximately between 900 nm and up to 1600 nm is reflected.[15] In addition, the photons in the high-energy spectral range where the cell response is

[15]The characteristics of the coating beyond the NIR range are not mentioned although they must play a role in the overall thermal performances via thermal radiation cooling to the cold outer space.

zero are also reflected (see Sect. 3.5.2). A simple thermal balance between the solar radiation power absorbed by the cell and not converted into electrical power, and the thermally radiated power, shows that the GaAs cell covered by a dual UV-NIR (blue-red) reflecting layer reduces the cell temperature by 14.7 K compared to the reference cell. By assuming a temperature coefficient of −2000 ppm K^{-1}, the gain in electrical power is estimated to be almost 3%.

Another solution to prevent sub-bandgap absorption is to perform a spectral splitting of the incident photons before they reach the PV device. In Maghanga et al. (2011), a Spectrally Selective Reflector (SSR) is added to the mirror that directs and concentrates light onto a c-Si cell. For a perfect SSR, it is assumed that all photons with energy below the bandgap are not reflected and those above the bandgap are fully reflected and thus sent toward the PV cell. As a result, in the chosen configuration where cooling is passive (with h = 5 W m^{-2} K^{-1}), the concentration factor is less than 10, and all photons not converted into electricity contribute to heating the cell, the cell temperature is lowered by 10 to 20 °C when using the perfect SSR (the gains in efficiency range from 0.8 to 1.8% assuming a temperature coefficient of −5000 ppm K^{-1}).

Low-energy photon filtering remains attractive for multi-junction cells although the heat source from thermalization is less than for single-junction cells as the fraction of incident radiation below the lowest bandgap is usually only a few percent of the total irradiation. In Geisz et al. (2010), different sub-bandgap filtering options are tested on inverted metamorphic triple-junction concentrator solar cells. The heat source can be lowered substantially (by 17–21%) compared to a base reference architecture. The arrangement that performs the best gets rid of the useless infrared light and thus could provide an additional 3.5% in conversion efficiency under 500 suns. Of course, the sub-bandgap filtering should not lower transmission on the useful range of the spectrum. In Micheli et al. (2012), different PMMA[16] layers are added on top of triple-junction concentrator solar cells to reflect radiation below the lowest bandgap of the multi-junction. Although the cells with these PMMA layers have a lower operating temperature, in the end they do not perform better than the reference cells because the coatings partially filter radiation above the bandgap as well. For such imperfect filters, there is a trade-off between reducing the heat load and increasing the optical losses. It depends on the operating conditions and the temperature coefficient of the device (see the analysis proposed in Sect. 3.5.2 in the case of high-energy photon filtering).

It is worth mentioning that filtering the sub-bandgap energy photons is beneficial only up to a wavelength of 2.5 μm for solar cells. Above this wavelength, the solar irradiance is negligible. It is thus better that the solar cell absorbs and thus emits above 2.5 μm as it allows for a radiative exchange between the cell and its environment. The radiative environment of the cell is typically the ground and the sky which are at temperatures not too different from ambient. In addition, there is a spectral windows in the atmosphere (8–13 μm) through which radiative exchange

[16]Polymethyl methacrylate.

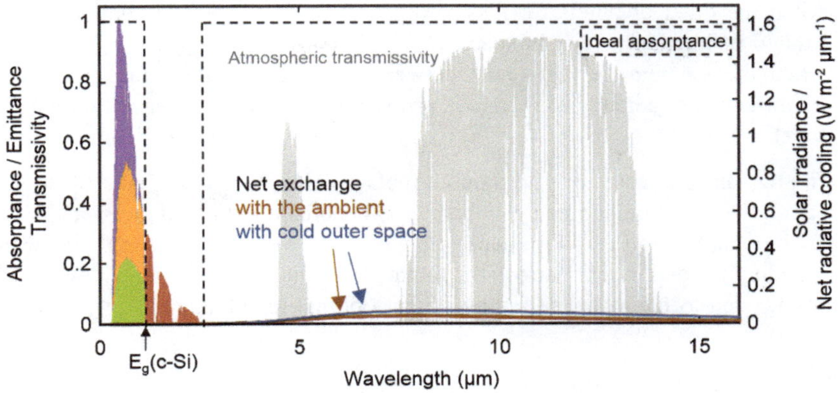

Fig. 3.6 Ideal absorptance profile to contribute to minimizing the operating temperature of a c-Si solar cell (*dashed line*). The AM1.5 solar irradiance is split in *purple*, *red*, *orange* and *green* areas which correspond to the heat generated by thermalization, the heat not generated via the reflection of sub-bandgap photons, the other heat sources associated with the PV conversion and the output power, respectively. The net exchanges with the ambient and cold outer space are shown in *brown* and *blue*. The exchange with cold outer space is only possible through the atmospheric windows shown in *grey*

with the cold outer space (~3 K) is possible (see Sect. 1.3.4). This spectral window is thus of particular importance for the radiative cooling of PV devices. When taking into account the operating temperature of a solar cell, an ideal absorptivity profile can be defined as shown in Fig. 3.6:

- the ideal cell should absorb perfectly (absorptance $A = 1$) all photons likely to generate electron-hole pairs (i.e. with more energy than the bandgap and conversely smaller wavelength). Note that for certain configurations, there might be a photon energy above which it is better to reflect the incoming radiation because the photogeneration comes at the expense of a too large heat source (see Sect. 3.5.2),
- above the wavelength corresponding to the cell bandgap and up to about 2.5 μm, the ideal cell should not absorb ($A = 0$) to prevent heat generation caused by the parasitic absorption of sub-bandgap photons,
- above about 2.5 μm, the cell should absorb perfectly ($A = 1$) in order to emit towards its environment (ground, sky and cold outer space) as much radiative energy as possible.

This ideal absorptance profile would contribute to minimize the cell operating temperature. The reduction of heat generated within the device due to the reflection of sub-bandgap photons is obviously a function of the bandgap of the absorber. For example it corresponds to 195 W m^{-2} for c-Si and 347 W m^{-2} for GaAs assuming a 1000 W m^{-2} AM1.5 illumination. Radiative cooling (determined by the absorptance above 2.5 μm and between 8 and 13 μm in particular) is especially important:

– when the other channels for dissipating heat (convection and conduction) are limited or inexistent (e.g. in space applications),
– when the temperature difference between the cell and the ambient becomes important (the radiative exchange is proportional to this difference to the power four).

Finally, in the case of thermophotovoltaic (TPV) devices (see Sect. 4.8), reflecting the sub-bandgap energy photons is beneficial both to the cell and to the thermal radiation source illuminating the cell. Indeed the radiation power not absorbed by the cell is sent back to the radiator and thus reduces the external power supply needed to maintain the radiator at an optimal temperature.

3.5.2 *High-Energy Photon Filtering*

Only the photons with energy larger than the bandgap are useful to the photo-conversion. In some cases, the spectral response of the PV device drops to zero above a high-energy cutoff. In Beauchamp et al. (1993), a UV reflection coating helps reducing the heat load, but marginally compared to the sub-bandgap filtering (Sect. 3.5.1), because the radiation power above the cell spectral response cutoff is only 5% of the total incident power for an AM0 illumination.

When not zero, the spectral response of solar cells at short wavelengths can still be poor. It can be interesting to down-shift these high-energy photons to lower energies by means of a luminescent layer deposited on the cell (Klampaftis et al. 2009). The primary impact is an increase of the number of incident photons that contribute to the photocurrent. The secondary improvement is the reduction of the thermalization heat source. Indeed, the high-energy photons carry more energy than needed for exciting an electron from the valence band to the conduction band. The energy in excess of the bandgap is converted into heat via thermalization (see Sect. 3.2). In Riekeberg et al. (2006), a mean to decouple thermalization from solar cells is explored. Thermalization is partially exported to a luminescent layer added on top of a cell. The layer is made of a direct bandgap semiconductor that absorbs the incident photons with energy larger than its bandgap ($E_{g,layer}$) while the sub-bandgap photons are transmitted directly to the cell. Photons converted by luminescence have thus an energy equal to the layer bandgap and are sent toward the cell that has a slightly smaller bandgap. With a GaAs layer ($E_{g,layer}$ = 1.42 eV) added onto a c-Si cell (E_g = 1.1 eV), in the ideal case,[17] 40% of the thermalization power that takes place in a regular c-Si cell could be exported to the luminescent

[17]Perfect absorption of the above-bandgap photons in the luminescent layer, 100% efficient luminescent conversion, perfect absorption of the luminescent photons in the cell, no reflection loss at the interfaces between layers. It is worth noticing that thermal management issues need to be addressed (i.e. evacuation of the residual heat generated in the cell and in the layer).

layer. In this configuration, a reduction of the thermalization heat source can be achieved without sacrificing photons that contribute to the photocurrent.

In STC, one design criterion is to reduce the optical losses, i.e. interband absorption within the cell should be maximized. But boosting absorption of the useful photons comes with increasing the thermalization heat source. This increases the equilibrium cell temperature in real operating conditions which results in a drop in conversion efficiency. In certain configurations, depending on the operating conditions, it might be better to reflect completely the highest energy photons than to absorb them, even though they would otherwise generate additional electrical carriers.

As a matter of fact, a photon energy cutoff (E_{cutoff}) above which absorbing photons is more detrimental than beneficial to the conversion efficiency can be derived (Dupré 2015). The driving parameters are the dependence of the total internal heat source to the cutoff energy, the ability of the device to evacuate heat (given by the heat transfer coefficient h and the ambient temperature T_{amb}) and the temperature dependence of conversion efficiency (given by the temperature coefficient β). Fig 3.7 shows the ratio of heat generated to useful power as a function of photon energy for a single-junction PV cell. The purple area corresponds to the fraction of the incident photon energy in excess of the cell bandgap that is thermalized. The limit is plotted as a straight line because this heat generation is a linearly increasing function of photon energy above bandgap. The red area corresponds to the fraction of the incident photon energy that results in heat generation from mechanisms other than thermalization. In this simplified example, it is assumed that recombinations do not depend upon the position in the cell where absorption takes place and thus do not depend on photon energy. By neglecting thermal radiation exchanges, the cell temperature results from the balance between the heat generated (Q) and that evacuated via convection:

$$T_c = T_{amb} + \frac{Q}{h} \tag{3.1}$$

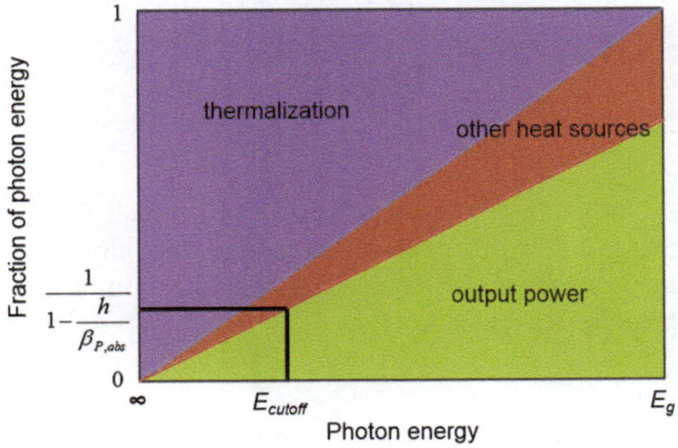

Fig. 3.7 Schematic illustrating the derivation of a photon high-energy cutoff in a simplified configuration. Figure from Dupré (2015)

In this configuration, an increase of the heat source (ΔQ) corresponds to an increase of the cell temperature:

$$\Delta T_c = \frac{\Delta Q}{h} \tag{3.2}$$

By introducing the cell absolute temperature coefficient:

$$\beta_{P,abs} = \frac{P(T_c) - P(T_{c,ref})}{T_c - T_{c,ref}} \tag{3.3}$$

an increase of the cell temperature implies a drop in efficiency[18]:

$$\Delta P = \beta_{P,abs}\, \Delta T_c \tag{3.4}$$

By combining these equations, an increase of the heat source leads to a power loss:

$$\Delta P = \beta_{P,abs}\, \frac{\Delta Q}{h} \tag{3.5}$$

It is possible to evaluate the fraction of heat generated to electrical power output above which photon absorption is detrimental to the cell efficiency. By noting the fraction of the incident photon energy E that results in heat ($F_Q(E)$) and the fraction that results in power output ($F_P(E)$), the condition for photon absorption to be beneficial reads:

$$F_P(E) + \beta_{P,abs}\, \frac{F_Q(E)}{h} > 0 \tag{3.6}$$

This simply means that the fraction of power output should be larger than the power drop that results from the cell temperature increase. Since $F_Q = 1 - F_P$ in this simplified configuration, the photon energy cutoff is the photon energy at which:

$$F_P(E_{cutoff}) = \frac{1}{1 - h/\beta_{P,abs}} \tag{3.7}$$

Although this result is derived in a simplistic configuration, it is meaningful in that it highlights the existence of a photon high energy cutoff and how it relates to the heat transfer coefficient (operating conditions and cell design) and to the temperature coefficient (cell design).

---

[18]The fundamental temperature coefficients of PV devices are negative (see Sect. 2.2).

In Bernardi et al. (2015), this principle is applied to derive a high-energy cutoff in the case of a near-field radiation thermophotovoltaic (NFR-TPV) device (see Sect. 4.8). A GaSb cell ($E_g = 0.723$ eV at 293 K) is illuminated by a thermal radiator at high temperature. In a given set of operating conditions, it is beneficial to filter out photons with energy above a cutoff (slightly above 1.1 eV), with an optimum (1.45 eV) that corresponds to a trade-off between sacrificing photocurrent and decreasing the thermalization heat source. At the optimum cutoff, a gain of 6. 5% in electrical power output is reached.

Reduction of the thermalization heat load is also possible by applying a partial filtering of the high-energy photons. An example of thermal optimization of the optimum thickness of an antireflection layer coated on a c-Si solar cell is presented in Sect. 3.5.4. When the real operating conditions are taken into consideration, the optimum thickness is such that there is more reflection of photons with the highest energies and thus less thermalization.

3.5.3 Bandgap Optimization

The bandgap of a photovoltaic device can be optimized as a function of the operating conditions, taking into account the thermal behavior of the device. The examples of solar cells (SCs) and solar thermophotovoltaic (STPV) systems are chosen by way of illustration.

In the case of solar cells, it is assumed for the sake of simplicity that all of the above bandgap photons are absorbed and excite an electrical carrier across the bandgap (i.e. no optical losses). All sub-bandgap photons are reflected as in Sect. 3.5.1, which means that the "sub-bandgap" loss does not contribute to the heat source. Except for the "sub-bandgap" and "emission" terms (Sect. 3.2), the incident power is converted into either heat or electrical power. Assuming that the Sun can be assimilated to a blackbody at 5800 K, Fig. 3.8 highlights that the heat generated in the cell depends on the bandgap. This heat source corresponds to the area bounded by the bold line. In the radiative limit (Fig. 3.8a, where the external radiative efficiency, ERE, is equal to 1), the larger the bandgap, the smaller the heat source, because thermalization decreases. When non-radiative recombinations take place (Fig. 3.8b where the ERE is equal to 10^{-5}), the terms that contribute to the heat source also decrease when the bandgap increases. In addition, the temperature coefficient of the cell conversion efficiency (dotted lines on Fig. 3.8), calculated using the methodology described in Chap. 2, is a decreasing function of the bandgap (even more so for lower EREs).

The joint decrease of the heat source and the temperature coefficient has an impact on the optimum bandgap as a function of the operating conditions. The larger the heat source, the larger the cell temperature for a given cooling configuration and in turn the smaller the conversion efficiency according to its temperature dependence.

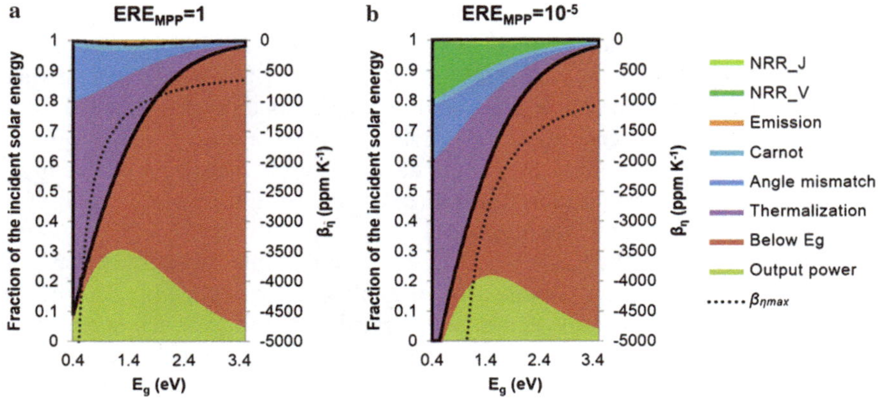

Fig. 3.8 Transformation of the incident solar radiation at the maximum power point at 300 K as a function of bandgap for a solar illumination modelled by a blackbody at 5800 K: **a** in the ideal case (*ERE* = 1, radiative limit), **b** for a realistic case (*ERE* = 10^{-5}, commercial cell). The losses contributing to the heat source are bounded by the *bold line*. The temperature coefficient of the conversion efficiency is shown by the *dotted line*. The color code of the losses is given on the *right*. Figure from Dupré (2015)

This is illustrated in Fig. 3.9 for a given set of operating conditions (heat transfer coefficient h = 10 W m^{-2} K^{-1}, ambient temperature T_{amb} = 25 °C, thermal radiation exchanges neglected). Similarly to the heat source, the cell temperature is a decreasing function of bandgap. The main result is that for both the ideal cell (*ERE* = 1) and the more realistic cell (*ERE* = 10^{-5}), the bandgap that gives the

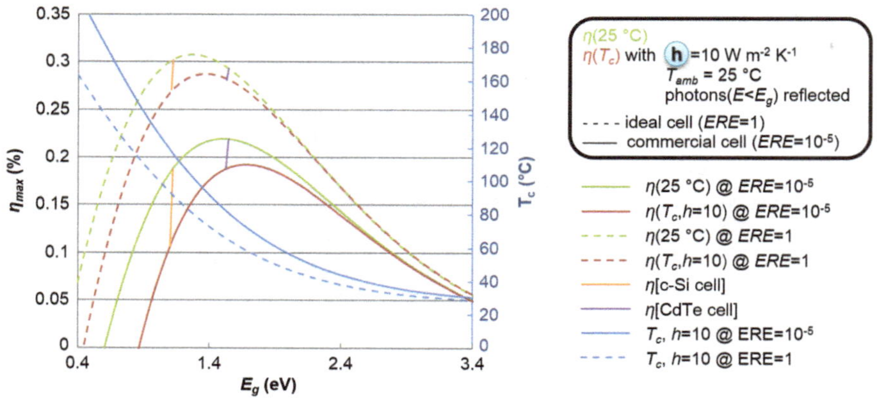

Fig. 3.9 Cell operating temperature and efficiency as a function of bandgap on for an ideal cell (*ERE* = 1, *dotted blue lines*) and a commercial cell (*ERE* = 10^{-5}, *full blue lines*) when h = 10 W m^{-2} K^{-1} and T_{amb} = 25°C. The efficiencies in *green* are for STCs and in *red* for the calculated operating temperatures. The efficiencies of a c-Si cell (*orange*) and a CdTe cell (*purple*) are indicated. Figure from Dupré (2015)

largest efficiency is different when the device temperature is assumed to be 25 °C or when real operating conditions are considered (i.e. the cell temperature is calculated as a function of the heat source and the heat transfer coefficient). The second main result is that the impact is larger for the non-ideal cell, as both the heat source and the temperature coefficient are larger in that case compared to the ideal case (see Fig. 3.8).

A meaningful example developed in Dupré et al. (2016a) consists in performing the same calculations for an AM1.5 irradiation and two cells having different bandgaps at 25 °C ($\sim$1.1 eV for c-Si and $\sim$1.5 eV for CdTe) and EREs such that these two cells have the same efficiency in standard conditions (at 25 °C). Assuming the same thermal conditions (h = 10 W m^{-2} K^{-1}, T_{amb} = 25 °C), the CdTe cell performs better than the c-Si cell (efficiency of 17.9% against 14.5%) because both the operating temperature and the temperature coefficient are smaller for the CdTe cell.

A second example of thermal optimization investigates the best composition of $Al_xGA_{1-x}As$ single-junction solar cells for missions to Mercury (López et al. 2016). The specific challenge for that space application is limiting the efficiency drop in a high-intensity high-temperature (HIHT) environment (with expected operating temperatures above 200 °C). In space applications, thermal radiation is the only channel for the cell to evacuate heat. Coupled resolution of the detailed balance equation in the radiative limit for the cell efficiency and the power balance for the cell equilibrium temperature, indicates that the latter depends on the bandgap of the semiconductor material of the cell. Indeed varying the bandgap has an impact on the absorption and emission currents, on the heat load and thus on the cell equilibrium temperature and subsequently on the electrical power output. The optimum bandgap results from trade-offs between optical losses and thermal losses (resulting from heat load and temperature coefficient changes). In this example, (López et al. 2016), two different bandgaps maximizing the electrical power output are found corresponding to two operating conditions (two altitudes of the space probe). When a sub-bandgap energy reflector is added (see Sect. 3.5.1), the optimum bandgaps are shifted to larger values which correspond to lower operating temperatures. The shifts are due to the reduction of the heat load caused by the reflection of the sub-bandgap photons. Thus, in space applications, particularly with harsh environments, the bandgap of solar cells must be selected carefully as a function of the operating conditions and taking into account design parameters such as sub-bandgap energy photon reflectors.

A third example is about the selection of the bandgap that maximizes the overall efficiency of solar thermophotovoltaic (STPV) devices. In this case, the thermal radiator illuminating the TPV cell is heated by the absorption of concentrated solar irradiation (see Sect. 4.8). In Dupré et al. (2016b), it is shown that the optimum bandgap of TPV cells depends on the performances of the cell cooling system (defined by a convective heat transfer coefficient), and on the quality of the cell (defined by its external radiative efficiency, see Sect. 2.3). In the simulations, a sub-bandgap filtering (see Sect. 3.5.1) is systematically considered. It has the dual advantage of reducing the heat load in the cell and of supplying power back to the thermal radiator. The optimum bandgaps result from trade-offs between optical, electrical and thermal losses. For example, for a given set of operating conditions, the optimal bandgap increases with decreasing ERE because heat sources and temperature coefficients increase with ERE.

The bandgap is thus a critical parameter governing the minimization of the thermal losses. Its optimum depends on the operating conditions, but also on other design parameters of PV devices.

3.5.4 Optimization of Other Parameters

The energy bandgap is not the sole characteristic of a photovoltaic device whose optimum might be different in real operating conditions than in STC. Optimizations of photovoltaic devices that include the design rule "minimizing the internal heat load" are illustrated in Dupré and Vaillon (2014). For the sake of simplicity, encapsulation layers are not considered in this article. Two design parameters of a bare single-junction crystalline silicon (c-Si) cell, the cell and the antireflection coating (ARC) thicknesses, are investigated as a function of thermal boundary conditions (variable heat transfer coefficient, prescribed ambient temperature). The optimum cell thickness decreases with the heat transfer coefficient as a result of larger cell temperatures caused by more internal heat generated from thermalization and recombinations. The optimum ARC thickness increases when the heat transfer coefficient decreases because the ARC thickness has an impact on the spectral distribution of the absorbed photons, and thus, on recombination and thermalization (see Sect. 3.5.2 about high-energy photon filtering). When the ARC thickness increases, so does the wavelength at which reflection is zero (and conversely for photon energy). If only an optical design rule is applied, then the optimum ARC thickness is that which minimizes the reflection of the incident potentially useful photons (Fig. 3.10). If the optical and electrical design rules ("maximizing the power output in STC") are considered, then the optimum ARC thickness is shifted to smaller values because the carriers generated by low energy photons are more likely to recombine because they are generated further from the p-n junction. If the thermal design rule is added, the optimum ARC thickness is shifted to larger values so as to reflect more of the high energy photons that contribute the most to the thermalization heat source.

Similar "thermal" design principles are applied in Vogt et al. (2015) for a module composed of encapsulated crystalline silicon cells with antireflection coatings and rear reflectors. A 3D ray tracing optical modeling is coupled with a 1D FEM (COMSOL) solver of the electrical and thermal diffusion equations. The ray-tracing optical simulations allow determining the optical losses in the different layers constituting the module. A specific calculation indicates that replacing the full-area rear metallization with a SiN_x mirror layer leads to a decrease of the operating temperature by 3.2 °C and in turn to an increase of the electrical power output. The same kind of analysis is reapplied in Vogt et al. (2016) where the

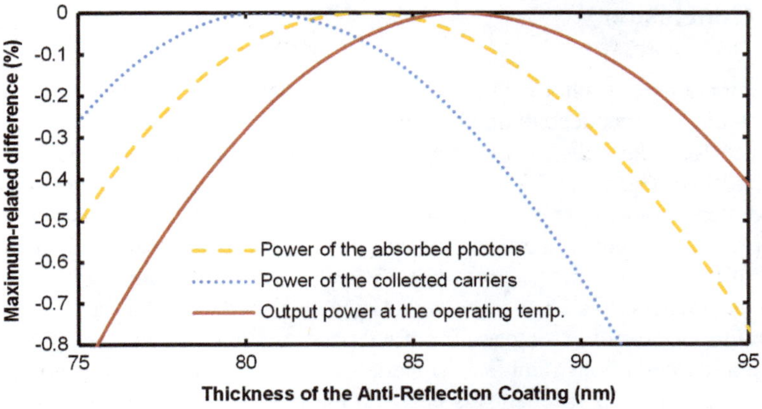

Fig. 3.10 Optimum antireflection coating thickness that minimizes optical losses, optical and electrical losses, optical-electrical-thermal losses for an operating condition different from STC. Figure created with data from Dupré and Vaillon (2014)

operating temperature of modules with p-type passivated emitter and rear cells (PERC) is reported to be lower than modules with cells that have a full-area rear aluminum metallization (Al-BSF), both from simulations and experiments.

The reason is that there is substantially more sub-bandgap radiation reflected by the PERC cell than by the Al-BSF cell (partial filtering of the sub-bandgap energy photons, see Sect. 3.5.1). The incident radiation power below the bandgap of c-Si is 192 W m^{-2} for the AM1.5 solar spectrum. Assuming that the reflectance of the PERC and Al-BSF type cells are respectively 40 and 25% in that spectral region, the thermal load below the bandgap is 29 W m^{-2} lower for the PERC cell. Taking also into account the differences in efficiencies of these two types of cells [η_{PERC}-$\eta_{Al\text{-}BSF}$ = 20.5% − 18.6% = 1.9%, from (Vogt et al. 2016)], there is an additional reduction in the internal heat source of 19 W m^{-2} for the PERC cell. In total, the PERC cell heat load is ΔQ = 47.8 W m^{-2} lower. Assuming that the cell evacuates heat at both sides through natural convection (with h = 5 W m^{-2} K^{-1}), then according to Eq. (3.2) the PERC cell temperature should be about $\Delta Q/(2\,h)$, i.e. 4.8 K, lower. This rough estimation is in good agreement with the value (4 K) experimentally measured in Vogt et al. (2016). In addition, it should be noticed that PERC cells are expected to have better temperature sensitivities because of their larger open-circuit voltages (see Sect. 2.3.1).

The optimizations of different cell parameters beyond STC are generally not independent from each other. In the examples of Sect. 3.5.3, the best bandgaps could be a function of considering a sub-bandgap reflector or not, or of the ERE of the cell. In the present section, selected examples have shown that other cell parameters can be tuned to lower the thermal load, but for a given bandgap. This means that there is room for optimizing other PV device parameters considering thermal design rules, one parameter at a time, and eventually all together.

3.6 Conclusion

The temperature of photovoltaic devices is usually several degree above the ambient temperature, depending on operating conditions (such as wind velocity and irradiance for solar cells), because a large fraction of the incident solar energy is converted into heat via different loss mechanisms. It is crucial to be able to predict the operating temperature of PV devices because their efficiency is function of their temperature (via temperature coefficients, see Chap. 2). In Sect. 3.2, a detailed model of the different heat sources in a PV cell, which highlights the effects of the device parameters, has been presented. Using this model for the heat source, the equilibrium or operating temperature of a device can be calculated as a function of the operating conditions (ambient temperature, wind speed, etc.) by evaluating the different heat transfer mechanisms between the device and its surroundings (conduction, convection and radiation). Similarly to the heat generated within the device, the heat transfer can depend upon certain device characteristics. For example, it is conceivable to increase convective heat transfer by adding fins to the back of a module.

Using this thermal model to obtain the device operating temperature highlights that the definition of the nominal operating cell temperature (NOCT) could be revisited. Indeed, the NOCT is measured on open-circuited modules while the heat source and thus the operating temperature is lower at the maximum power point (MPP) because a fraction of the incident power is converted into electricity instead of heat. This suggests that a NOCT defined at the MPP would represent more accurately the actual operating temperature of a device. Also, since thermal boundary conditions are highly dependent on location and on mounting configuration, it could be useful to define a set of NOCTs for different representative operating conditions.

In the standard test conditions (STC), performances of PV devices are evaluated on their *ability to absorb incident radiation for generating electrical charges* and *to collect these charges at* 25 °C. In real operating conditions, a third performance criterion must be added: their *ability to mitigate the thermal losses*. A PV device that performs better than another one with respect to the first two criteria in STC, may underperform in real operating conditions. Indeed, when the cell device operating temperature is calculated using a thermal model rather than assuming a fixed temperature, the operating efficiency becomes a function of the heat source, the cooling of the device and its temperature sensitivity. Thus, the "thermal" design approach consists in maximizing the cooling, minimizing the heat source, and minimizing the temperature coefficient in the optimization of all the device parameters. This approach is proven successful for different parameters (such as bandgap, anti-reflection coating thickness or sub-bandgap reflectivity) and for different systems, from standard PV cells to solar and near-field thermophotovoltaic devices. The efficiency improvements from these optimizations due to reductions in operating temperature usually come at no added cost in the device fabrication. Furthermore, because the effects of all the parameters are deeply intertwined, larger benefits are expected from a global optimization.

References

Atwater H, Polman A (2010) Plasmonics for improved photovoltaic devices. Nat Mater 9:205–213. Doi:10.1038/nmat2629

Barnett A, Kirkpatrick D, Honsberg C et al (2009) Very high efficiency solar cell modules. Prog Photovoltaics Res Appl 17:75–83. Doi:10.1002/pip.852

Beauchamp WT, Tuttle-Hart T (1995) UV/IR reflecting covers for solar cells (patent US 5449413 A)

Beauchamp WT, Tuttle-Hart T, Sanders ML (1993) Blue/red reflecting solar cell coversfor GaAs cells. In: Proceedings of the 23rd IEEE PVSC. pp 1487–1490

Bergman TL, Lavine AS, Incropera FP, DeWitt DP (2011) Fundamentals of heat and mass transfer, 7th edn. Wiley, New York

Bernardi MP, Dupré O, Blandre E et al (2015) Impacts of propagating, frustrated and surface modes on radiative, electrical and thermal losses in nanoscale-gap thermophotovoltaic power generators. Sci Rep. Doi:10.1038/srep11626

Breitenstein O, Rakotoniaina JP (2005) Electrothermal simulation of a defect in a solar cell. J Appl Phys. Doi 10(1063/1):1866474

Chen D (2001) Antireflection (AR) coatings made by sol-gel processes: a review. Sol Energy Mater Sol Cells 68:313–336

Chen G (2005) Nanoscale energy transport and conversion. Oxford University Press, Oxford

Dupré O (2015) Physics of the thermal behavior of photovoltaic devices. INSA Lyon, France

Dupré O, Vaillon R (2014) Optimizations of photovoltaic cells including the minimization of internal heat sources. J Renew Sustain Energy 6:11201. Doi:10.1063/1.4828367

Dupré O, Vaillon R, Green MA (2015) Physics of the temperature coefficients of solar cells. Sol Energy Mater Sol Cells 140:92–100. Doi:10.1016/j.solmat.2015.03.025

Dupré O, Vaillon R, Green MA (2016a) A full thermal model for photovoltaic devices. Sol Energy 140:73–82. Doi: 10.1016/j.solener.2016.10.033

Dupré O, Vaillon R, Green MA (2016b) Optimization of solar thermophotovoltaic systems including the thermal balance. In: Proceedings of the 43rd IEEE PVSC

Geisz JF, Ward JS, Duda A et al (2010) Infrared reflective and transparent inverted metamorphic triple junction solar cells. AIP Conf Proc 1277:11–15. Doi:10.1063/1.3509169

Green MA (1984) Limits on the open-circuit voltage and efficiency. IEEE Trans Electron Devices ED-31:671–678

Green MA (2003) Third generation photovoltaics. Advanced solar energy conversion. Springer, Berlin

Green MA (2012) Radiative efficiency of state of the art photovoltaic cells. Prog Photovoltaics Res Appl 20:472–476. Doi:10.1002/pip

Green MA, Ho-Baillie A (2010) Forty three per cent composite split-spectrum concentrator solar cell efficiency. Prog Photovoltaics Res Appl 18:42–47. Doi:10.1002/pip.924

Harder N, Würfel P (2003) Theoretical limits of thermophotovoltaic solar energy conversion. Semicond Sci Technol 151:S151–S157. Doi:10.1088/0268-1242/18/5/303

Heikkilä O, Oksanen J, Tulkki J (2009) Ultimate limit and temperature dependency of light-emitting diode efficiency. J Appl Phys. Doi 10(1063/1):3125514

Hirst LC, Ekins-daukes NJ (2011) Fundamental losses in solar cells. Prog Photovoltaics Res Appl 19:286–293. Doi:10.1002/pip.1024

Holst H, Winter M, Vogt MR et al (2013) Application of a new ray tracing framework to the analysis of extended regions in Si solar cell modules. Energy Procedia 38:86–93. Doi:10.1016/j.egypro.2013.07.253

Howell JR, Mengüç MP, Siegel R (2015) Thermal radiation heat transfer, 6th edn. CRC Press, Boca Raton

Jackson E (1955) Areas for improvement of the semiconductor solar energy converter. In: Transactions of the conference on the use of solar energy. Tucson, Arizona, p 5

Jones AD, Underwood CP (2001) A thermal model for photovoltaic systems. Sol Energy 70:349–359. Doi:10.1016/S0038-092X(00)00149-3

King RR, Sherif RA, Kinsey GS, et al (2005) Bandgap Engineering in high-efficiency multijunction concentrator cells. In: Proceedings of the international conference on solar concentrators for the generation of electricity or hydrogen

King RR, Bhusari D, Boca A et al (2011) Band gap-voltage offset and energy production in next-generation multijunction solar cells. Prog Photovoltaics Res Appl 19:797–812

Kittel C (1996) Introduction to solid state physics, 7th edn. Wiley, New York

Klampaftis E, Ross D, McIntosh KR, Richards BS (2009) Enhancing the performance of solar cells via luminescent down-shifting of the incident spectrum: a review. Sol Energy Mater Sol Cells 93:1182–1194. Doi:10.1016/j.solmat.2009.02.020

Koehl M, Heck M, Wiesmeier S, Wirth J (2011) Modeling of the nominal operating cell temperature based on outdoor weathering. Sol Energy Mater Sol Cells 95:1638–1646. Doi:10.1016/j.solmat.2011.01.020

König D, Casalenuovo K, Takeda Y et al (2010) Hot carrier solar cells: Principles, materials and design. Phys E Low-Dimensional Syst Nanostruct 42:2862–2866. Doi:10.1016/j.physe.2009.12.032

Kosten EDE, Atwater JHJ, Parsons J et al (2013) Highly efficient GaAs solar cells by limiting light emission angle. Light Sci Appl 2:1–6. Doi:10.1038/lsa.2013.1

López E, Martí A, Llorens JM, et al (2016) Optimum single-gap solar cells for missions to Mercury. J Spacecr Rockets, 1–5. Doi:10.2514/1.A33533

Luque A, Martí A, Stanley C (2012) Understanding intermediate-band solar cells. Nat Photonics 6:146–152. Doi:10.1038/nphoton.2012.1

Maghanga CM, Niklasson G a, Granqvist CG, Mwamburi M (2011) Spectrally selective reflector surfaces for heat reduction in concentrator solar cells: modeling and applications of TiO_2: Nb-based thin films. Appl Opt 50:3296–302. Doi:10.1364/AO.50.003296

Martí A, Araújo GL (1996) Limiting efficiencies for photovoltaic energy conversion in multigap systems. Sol Energy Mater Sol Cells 43:203–222. Doi:10.1016/0927-0248(96)00015-3

Martí A, Balenzategui JL, Reyna RF (1997) Photon recycling and Shockley's diode equation. J Appl Phys 82:4067–4075. Doi:10.1063/1.365717

Micheli L, Sarmah N, Luo X, et al (2012) Infrared reflecting coverglasses for multijunction cells in a terrestrial high-concentrating photovoltaic system. In: 27th European photovoltaic solar energy conference and exhibition. pp 266–270

Miller OD, Kurtz SR, Yablonovitch E (2012) Strong Internal and External Luminescence as solar cells approach the Shockley-Queisser limit. IEEE J Photovoltaics 2:303–311. Doi:10.1109/JPHOTOV.2012.2198434

Modest MF (2013) Radiative heat transfer, 3rd edn. Academic Press, New York

Pillai S, Green MA (2010) Plasmonics for photovoltaic applications. Sol Energy Mater Sol Cells 94:1481–1486. Doi:10.1016/j.solmat.2010.02.046

Pipe K, Ram R, Shakouri A (2002) Bias-dependent Peltier coefficient and internal cooling in bipolar devices. Phys Rev B 66:1–11. Doi:10.1103/PhysRevB.66.125316

Pop E (2010) Energy dissipation and transport in nanoscale devices. Nano Res 3:147–169. Doi:10.1007/s12274-010-1019-z

Richter A, Hermle M, Glunz SW (2013) Reassessment of the limiting efficiency for crystalline silicon solar cells. IEEE J Photovoltaics 3:1184–1191. Doi:10.1109/JPHOTOV.2013.2270351

Riekeberg S, Altermatt PP, Brendel R (2006) Decoupling thermalisation from solar cells. In: Proceedings of the 21st EUPVSEC. Dresden, Germany, pp 294–297

Ross R, Gonzalez C (1980) Reference conditions for reporting terrestrial photovoltaic performance. In: AS/ISES Annual Meeting. pp 1091–1097

Ross RT, Nozik AJ (1982) Efficiency of hot-carrier solar energy converters. J Appl Phys 53:3813–3818. Doi:10.1063/1.331124

Ross RG, Smokler MI (1986) Electricity from photovoltaic solar cells: flat-plate solar array project final report. Volume VI: engineering sciences and reliability

Silva JP, Nofuentes G, Munoz JV (2010) Spectral reflectance patterns of photovoltaic modules and their thermal effects. J Sol Energy Eng Asme 132:13. Doi:10.1115/1.4002246

Trupke T, Green MA, Würfel P (2002a) Improving solar cell efficiencies by up-conversion of sub-band-gap light. J Appl Phys 92:4117–4122. Doi:10.1063/1.1505677

Trupke T, Green MA, Würfel P (2002b) Improving solar cell efficiencies by down-conversion of high-energy photons. J Appl Phys 92:1668–1674. Doi:10.1063/1.1492021

Vogt MR, Holst H, Winter M et al (2015) Numerical modeling of c-Si PV modules by coupling the semiconductor with the thermal conduction, convection and radiation equations. Energy Procedia 77:215–224. Doi:10.1016/j.egypro.2015.07.030

Vogt MR, Schulte-huxel H, Offer M, et al (2016) Reduced module operating temperature and increased yield of modules with PERC instead of Al-BSF solar cells. In: 43rd IEEE photovoltaic specialist conference. pp 1–6

Volz S (2007) Microscale and nanoscale heat transfer. Topics in applied physics 107. Springer. Doi: 10.1007/11767862

Würfel P (1997) Solar energy conversion with hot electrons from impact ionisation. Sol Energy Mater Sol Cells 46:43–52. Doi:10.1016/S0927-0248(96)00092-X

Würfel P (2009) Physics of solar cells: from basic principles to advanced concepts. Wiley. ISBN: 978-3-527-40857-3

Xue J, Zhao Y, Oh SH et al (2015) Thermally enhanced blue light-emitting diode. Appl Phys Lett 10(1063/1):4931365

Zhang ZM (2007) Nano/microscale heat tarnsfer. McGraw-Hill, New York City

Chapter 4
Specificities of the Thermal Behavior of Current and Emerging Photovoltaic Technologies

Abstract This chapter describes the specificities of the thermal behavior of different photovoltaic technologies. In standard silicon solar cells, the temperature coefficients are mainly driven by the open-circuit voltage. It is argued that the upcoming deployment of the passivated emitter and rear cell (PERC) technology will be beneficial for the temperature sensitivities and the operating temperatures of modules. Silicon heterojunction (SHJ) solar cells also benefit from large open-circuit voltages which result in advantageous temperature coefficients. The thermal behavior of SHJ devices is also driven by specific mechanisms such as thermionic carrier transport which results in non-linear temperature variations of the fill factor and subsequently of the conversion efficiency. An analysis of the temperature sensitivity of solar cells made of compensated silicon is presented. The specificities of the short-term thermal behavior of a-Si solar cells are discussed. A review of the literature investigating the thermal behavior of perovskite-based devices is proposed. Despite the discrepancies, several important mechanisms are highlighted (e.g. the strong temperature dependence of carrier transport which generally results in a maximum of the conversion efficiency around room temperature). The particularities in terms of thermal behavior of multi-junction solar cells and concentrator devices are discussed. Finally, the analysis is extended to thermophotovoltaic applications where the operating conditions are significantly different from "standard" photovoltaics (different heat sources and angle mismatch, opportunity of low-energy photon recycling, etc.).

4.1 Standard Silicon Solar Cells

The temperature sensitivity of standard (i.e. p-n homo-junction) c-Si solar modules on the market in 2016 is typically around $-0.45\%\ K^{-1}$. This is significantly more than all the other PV technologies as shown in Fig. 4.1.

In general, the thermal behavior of c-Si solar cells is well described by the theoretical framework of Chap. 2. In summary:

O. Dupré et al., *Thermal Behavior of Photovoltaic Devices*,
DOI 10.1007/978-3-319-49457-9_4

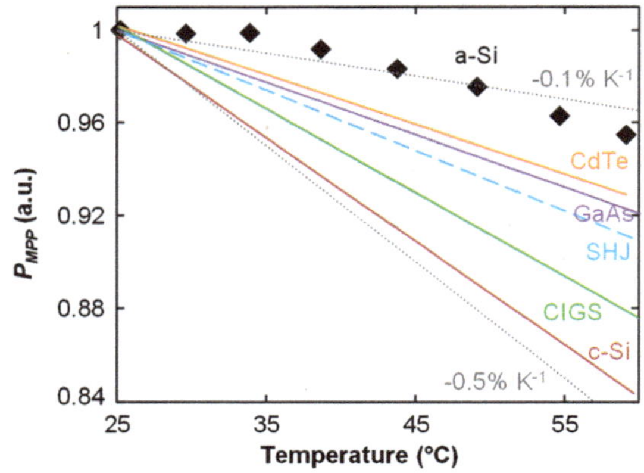

Fig. 4.1 Maximum power output as a function of device temperature of PV cells made of different semiconductors

- the short-circuit current density increases linearly with temperature. This originates from two phenomena: the negative temperature sensitivity of the bandgap of silicon and the positive temperature sensitivity of its inter-band absorption coefficient due to its indirect bandgap (see Sect. 2.3.2). Because of this second effect, cells with low external quantum efficiency (EQE) tend to have larger (positive) temperature coefficients of short-circuit current (Dupré et al. 2015),
- the open-circuit voltage decreases linearly with temperature. This temperature dependence is a function of the dominant recombination mechanisms but generally decreases as open-circuit voltage increases (see Sect. 2.3.1),
- the fill factor has also a negative temperature dependence which is ideally closely linked to that of the open-circuit voltage but also depends on transport mechanisms when these are limiting factors (see Sect. 2.3.3).

The open-circuit voltage is usually the main driver of the overall temperature sensitivity of c-Si solar cells. As a result, high open-circuit voltages are concomitant with advantageous temperature sensitivity (see Fig. 4.2). This means that the deployment of advanced type of solar cells (with larger V_{oc}) such as "passivated emitter and rear cells" (PERC) will enable smaller temperature coefficients and thus better performance ratios (Green, Sect. 1 in Boriskina et al. 2016). Ultimately, in the intrinsic limit dictated by Auger recombinations, c-Si solar cells could reach an open-circuit voltage of 761 mV and operate at 29.4% (Richter et al. 2013) with a temperature coefficient of −2380 ppm K^{-1} (Dupré 2015).

In addition to better temperature coefficients, lower operating temperatures are expected from more efficient devices because the energy which is converted into electricity does not contribute to heating the devices.[1] For example, the improvement in efficiency and the increase of sub-bandgap photon reflection result in a reduction of about 4 °C in operating temperature between PERC cells and Al-BSF

[1]This means that more efficient devices will also show larger energy yields (defined as the ratio between the energy produced in kWh divided by the rated power in kW).

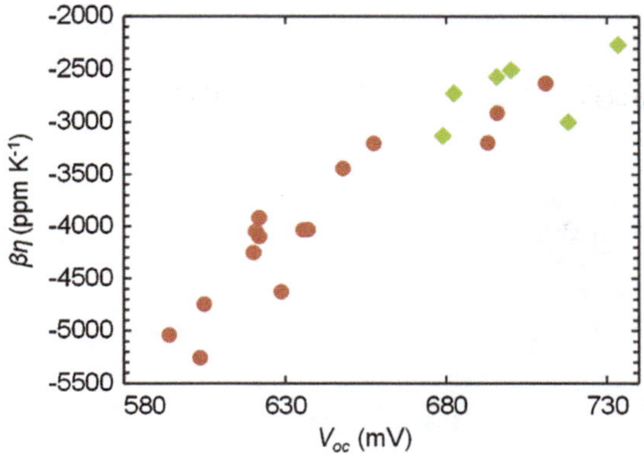

Fig. 4.2 Temperature coefficient of conversion efficiency as a function of open-circuit voltage. The *circles* correspond to homo-junction c-Si cells (Green et al. 1982, 1985; Zhao et al. 1994; Mishima et al. 2011; Tanay et al. 2011; Ponce-Alcántara et al. 2014; Xiao et al. 2014) and the diamonds correspond to SHJ cells (Mishima et al. 2011; Seif et al. 2014)

(Aluminium back surface field) cells (see Sect. 3.5.4). Note that additional reductions in operating temperature could be achieved by considering the thermal behavior of the devices in the optimization of their design parameters (see Chap. 3).

4.2 Silicon Hetero-Junction Solar Cells

Silicon hetero-junction (SHJ) or HIT[2] solar cells are known for their good energy yield originating from small temperature coefficients (around $-0.25\%\ K^{-1}$ as compared to about $-0.45\%\ K^{-1}$ for standard c-Si cells). The large open-circuit voltage achieved with this cell structure[3] is the main reason for these favorable temperature coefficients. Figure 4.2 illustrates the concomitant advantageous temperature coefficient and high open-circuit voltage of SHJ cells (depicted by diamonds) in comparison with standard homo-junction c-Si cells (depicted by circles). The relation between β_η and V_{oc} originates from the fact that the rates of all the recombination mechanisms (SRH, Auger, radiative, surface) generally increase with temperature (see Sect. 2.3.1). Thus, the efficiency of a cell with a small

[2]HIT stands for Hetero-junction with Intrinsic Thin layers.

[3]SHJ solar cells generally consist of a layer of c-Si (being the absorber) sandwiched between two very thin layers (few nanometers) of amorphous silicon (a-Si) with different doping (that serve as selective membranes). These layers are separated by very thin layers of excellent quality amorphous silicon that passivate the interfaces (Mishima et al. 2011; De Wolf et al. 2012).

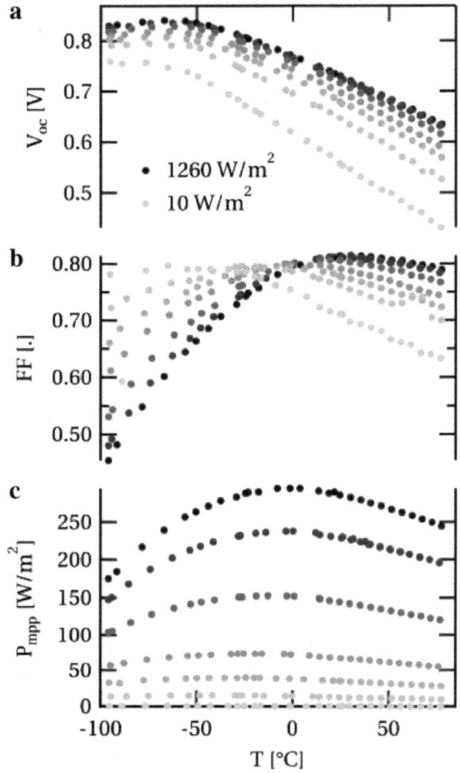

Fig. 4.3 Measurements on a silicon heterojunction solar cell of: **a** open-circuit voltage, **b** fill factor, **c** power at the maximum power point, as a function of temperature and for various illumination levels (from 10 W m^{-2} in *light grey* up to 1260 W m^{-2} in *black*). Figure from Riesen (2016)

recombination rate at 25 °C (large V_{oc}) decreases less when its temperature rises than that of a cell with a larger recombination rate (smaller V_{oc}).

In Fig. 4.2, the trend of SHJ cells seems quite similar to that followed by standard homo-junction silicon cells. However, there are some differences in the thermal behavior of the two types of device.

Figure 4.3a shows the open-circuit voltage of a SHJ solar cell as a function of temperature for different irradiances. Above a certain temperature (e.g. −30 °C for an illumination of approximately 1000 W m^{-2}), the open-circuit voltage decreases linearly with increasing temperature as for most PV devices (see Sect. 2.3.1). However, below this temperature, the open-circuit voltage exhibits an unusual behavior: it saturates and then starts decreasing with decreasing temperature. This behavior is correlated with the apparition of a s-shape in the I–V characteristic of the device (see Fig. 4.4). This s-shape can be attributed to a valence band offset between the c-Si absorber and the a-Si:H layer at the p-side of the device which impedes the collection of holes (Van Cleef et al. 1998; Riesen 2016). The s-shape in the I–V characteristic disappears above a certain temperature due to increasing

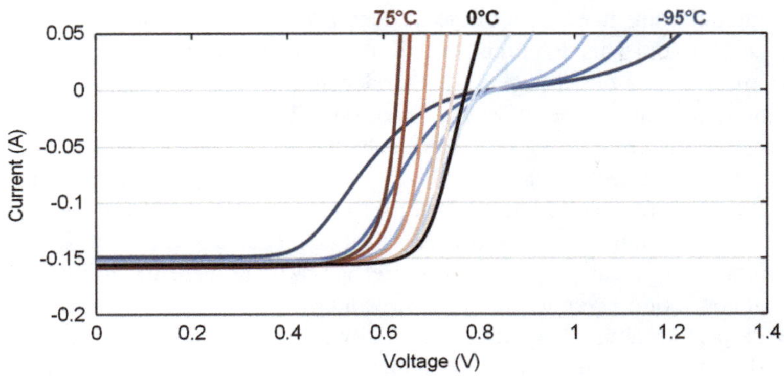

Fig. 4.4 Current-voltage (I–V) characteristics of a SHJ solar cell on a range of temperatures from −95 to +75 °C. Figure created with data from Riesen (2016)

thermionic emission[4] of holes across the transport barrier caused by the valence band offset.

The temperature at which this transport barrier starts to limit carrier transport also depends on the current and thus on the illumination level (see Fig. 4.3a). At low illumination levels, the open-circuit voltage saturation begins at low temperatures because the current is small which reduces the accumulation of holes and thus the modification of the field at the interface (Van Cleef et al. 1998).

Beyond its impact on the open-circuit voltage, this transport barrier has a similar effect on the fill factor as illustrated in Fig. 4.3b. It causes the fill factor to reach a maximum at a certain temperature. For the same reasons that those detailed for the open-circuit voltage, this temperature is a function of the illumination level. This maximum of the fill factor is the result of two competing phenomena:

- a reduction of the fill factor associated with the decrease of open-circuit voltage with increasing temperature due to increasing recombination rates (see Sect. 2.3.3),
- an augmentation of the fill factor resulting from the increase of thermionic emission of holes across the transport barrier created by the valence band offset.

This second effect explains the advantageous temperature coefficient of the fill factor of SHJ cells noticed in the literature (Taguchi et al. 2008; Seif et al. 2014, 2015; Riesen 2016).

It is interesting to note that the maximum of the fill factor can be in the range of usual PV cell operating temperatures especially at high illumination levels (e.g. ~25 °C at 1260 W m^{-2} for the cell considered in Fig. 4.3). This means that, for certain SHJ devices, a single temperature coefficient cannot represent accurately the temperature dependence of the fill factor.

---

[4]Thermionic emission increases with increasing temperature since it corresponds to the transport of carriers whose thermal energy is sufficient to cross the barrier (Sze and Ng 1981).

Additionally, this maximum of the fill factor results in a maximum of the output power (Fig. 4.3c). This maximum takes place at a lower temperature than the maximum of the fill factor because it also depends on the temperature sensitivities of open-circuit voltage and short-circuit current. This specific temperature dependence of the performances of SHJ solar cells has important consequences on their optimum design. Indeed, it has been shown experimentally that different parameters such as the thickness and doping of the a-Si:H layers have an important impact on the temperature coefficients of the devices (Taguchi et al. 2008; Seif et al. 2015). An interesting example can be found in Seif et al. (2014) where the use of an amorphous silicon oxide layer results in an advantageous β_{FF} and β_{η} which translates into better performances than the reference device at temperatures above 45 °C (see Fig. 2.20). This result illustrates the potential of optimizing the design parameters of SHJ devices as a function of specific operating conditions (temperatures and illuminations).

4.3 Compensated Silicon Solar Cells

The temperature sensitivities of cells made of compensated silicon, also-called upgraded metallurgical silicon (UMG-Si) or solar-grade silicon directly purified by metallurgical route (SoGM-Si), are investigated in several articles (Tanay et al. 2011; Tayyib et al. 2013, 2014a, b; Ponce-Alcántara et al. 2014; Xiao et al. 2014). This kind of silicon shows some promises for the photovoltaic industry because it is cheaper to produce than the electronic-grade silicon (EG-Si) typically used in PV cells. One particularity of this kind of silicon feedstock is that it contains high concentrations of Boron (B) and Phosphorous (P) that are respectively p-type and n-type dopants. This is the reason why it is named "compensated silicon". Also, compensated silicon feedstock contains much larger amounts of metal impurities and light elements (oxygen, carbon) than standard EG-Si (Tanay et al. 2011).

In the literature, it is argued that solar cells made of compensated silicon benefit from advantageous temperature coefficients of their efficiencies (β_{η}). The experiments show that these favorable β_{η} originate from larger β_{Jsc} which are generally attributed to smaller (compared to EG-Si) decreases with temperature of the minority carriers mobility. Indeed, mobilities in UMG-Si are more impacted by impurity scattering which is temperature independent and thus less driven by phonon scattering which increases with temperature (Xiao et al. 2014). Another reason suggested for the larger β_{Jsc} of cells made of compensated silicon feedstocks is that B-O defects are the main cause of recombinations in these cells and the lifetimes corresponding to these defects are known to increase significantly with temperature (Tanay et al. 2011).

Figure 4.5 shows the data from three independent studies on the temperature sensitivities of compensated silicon cells (Tanay et al. 2011; Ponce-Alcántara et al. 2014; Xiao et al. 2014). It appears that most of the measured cells made of compensated silicon have smaller efficiencies than the cells made of standard silicon

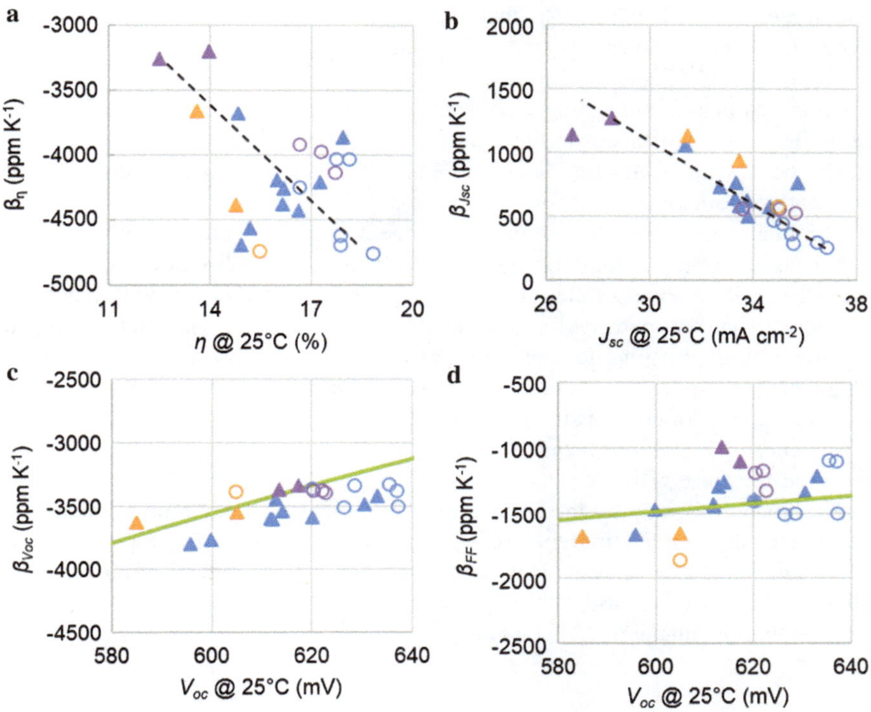

Fig. 4.5 Temperature coefficients (TCs) of solar cells made of standard EG-Si (*open circles*) and compensated Si (*filled triangles*). The data in *red*, *blue* and *orange* are respectively from Tanay et al. (2011), Ponce-Alcántara et al. (2014), Xiao et al. (2014). *Dashed lines* are guides to the eye to visualize the trends in the data. **a** TC of efficiency density plotted against efficiency at 25 °C; **b** TC of short-circuit current density plotted against short-circuit current density at 25 °C; **c** TC of open-circuit voltage plotted against open-circuit voltage at 25 °C. The *green line* corresponds to the theoretical expression (2.30) with $\gamma = 1.2$; **d** TC of the fill factor plotted against open-circuit voltage at 25 °C. The *green line* corresponds to the theoretical expression (2.55) with $\beta_{Rs} = 6450$ ppm K^{-1} and $r_s = R_s/V_{oc}/I_{sc} = 0.06$

(see Fig. 4.5a). This suggests that the advantageous temperature coefficients of cells made of compensated silicon come at the price of reduced performances at 25 °C. Indeed, while impurity scattering reduces the relative decrease of mobility with temperature, it also causes a reduction of the mobilities. Similarly, the recombination rate related to B-O defects decreases with temperature but it also reduces the global minority carrier lifetime compared to a cell without these defects.

In Fig. 4.5a, the different results seem to exhibit a similar trend: the temperature coefficient of efficiency becomes worse as efficiency improves. This is somewhat surprising as β_{Voc} usually accounts for a large part of β_η and it improves together with open-circuit voltage (see Sect. 2.3.1). Figure 4.5c shows that the relation between β_{Voc} and V_{oc} is similar for cells made of compensated and "standard" silicon feedstocks. This suggests that the temperature dependence of the dominant recombination mechanisms is similar in both type of cells.

The measured temperature coefficients of the fill factor, shown in Fig. 4.5d as a function of open-circuit voltage, are somewhat scattered. This is probably due to dissimilar series resistances in the different devices (see Sect. 2.3.3). Anyhow, the graph does not show any significant difference between cells made of compensated or "standard" silicon feedstocks.

The main reason why the temperature coefficient of efficiency decreases with increasing efficiency is the temperature coefficient of short-circuit current density. Indeed, Fig. 4.5b shows a clear trend: β_{Jsc} decreases significantly with increasing short-circuit current density for both kinds of silicon cells. This trend could be explained by the positive temperature dependence of the absorption coefficient of silicon originating from its indirect bandgap (see Sect. 2.3.2). The other phenomena described above for cells made of compensated silicon (i.e. the positive temperature dependences of mobilities and lifetimes) do not seem to modify the tendency observed for these cells.

The trends followed by the temperature coefficients of open-circuit voltage, fill factor and short-circuit current are similar for cells made of compensated and "standard" silicon feedstocks. This suggests that, if the efficiencies of compensated silicon solar cells increase, their temperature coefficients are likely to converge towards that of "standard" silicon cells.

4.4 Amorphous Silicon Solar Cells

Amorphous silicon (a-Si) solar cells have particularly good performances at elevated temperatures. Their thermal behavior is influenced by different phenomena acting at different time scales. One mechanism specific to a-Si is known as the Staebler-Wronsky effect (see Sect. 1.1.2). It corresponds to a metastable light-induced degradation which progressively increases defect concentrations and reduces the performances of a-Si devices. Fortunately, these light-induced changes can be reversed by thermal annealing. This temperature-promoted regeneration has a strong influence on the seasonal variations of the performances of a-Si PV modules (Makrides et al. 2012b). Indeed, high operating temperatures of modules in summer time enable an important regeneration. This is the reason why, in the field, positive temperature coefficients are sometimes reported for amorphous silicon modules (Fischer et al. 2009; Moser et al. 2013; Silverman et al. 2014). Also, this phenomenon explains that a-Si modules can have better energy yields with thermally insulated backsides, compared to a classic mounting configuration on open-racks (Kondo et al. 1997; Fischer et al. 2009).

The short-term thermal behavior of a-Si solar cells also presents some peculiarities. The temperature coefficients of the efficiency are usually in the range $[-0.1, -0.2]\%\ K^{-1}$ (Virtuani et al. 2010; Makrides et al. 2012a; Riesen 2016). These good temperature coefficients can be chiefly attributed to the large bandgap of amorphous silicon (~1.7 eV). As in c-Si solar cells, the short-circuit current increases with temperature because the bandgap of a-Si also decreases with temperature which leads to increased absorption in the infrared as temperature rises. In addition, for cells with high defect densities (and thus low fill factors), the short-circuit current temperature dependence is improved by the positive temperature sensitivity of carrier mobility (Riesen et al. 2016). As for the open-circuit voltage, its negative temperature sensitivity follows the expressions detailed in Sect. 2.3.1. The results in Riesen et al. (2016) describe the impact of different design parameters on the limiting loss mechanisms and thus on the temperature coefficients. The temperature dependence of the fill factor of a-Si cells is quite different from that of standard c-Si devices. As in SHJ cells (see Sect. 4.2), the fill factor has a maximum at a certain temperature (usually in the range of operating temperatures of PV devices) below which it drops with decreasing temperature. This effect is attributed to an accumulation of holes at the p-side interface at temperatures where the photogenerated hole concentration becomes significant compared to the concentration of holes thermally activated (Riesen et al. 2016). This explanation is consistent with the fact that the temperature of the maximum fill factor increases with increasing illumination level (Riesen et al. 2016). This behavior of the fill factor results in a maximum in efficiency as a function of device temperature (shifted to a lower temperature because of the negative temperature sensitivity of open-circuit voltage).

The unusual thermal behavior of a-Si devices implies that a device optimized at 25 °C does not necessarily give the best energy yield in actual operating conditions (Carlson et al. 2000; Riesen 2016). For example, it is suggested in Carlson et al. (2000) to increase the i-layer thickness for devices designed for PV roofing modules which are known to operate at particularly large temperatures. In addition, to optimize the i-layer thickness of a-Si devices for real operating conditions, one also needs to take into account illumination levels (Riesen et al. 2016). This is illustrated in Fig. 4.6 which shows the efficiency as a function of temperature of two cells with different i-layer thicknesses at three illumination levels. Because of the physics discussed previously, the efficiencies of the two devices cross each other at a temperature that depends on the illumination level. This means that the optimal design is not the same for two locations with similar operating temperatures but different irradiances. These studies pave the way towards the optimization of devices in real operating conditions (instead of the standard test conditions, see Sect. 3.5).

4.5 Perovskite Solar Cells

Perovskites are materials described by the formula ABX_3, where A and B are cations of different sizes (A larger than B) and X is an anion. A new class of solar cells based on mixed organic-inorganic halide perovskites (such as methyl-ammonium lead tri-iodide: $CH_3NH_3PbI_3$) has recently emerged and has been developing extraordinarily quickly (Green et al. 2014). The record efficiencies in standard test conditions

Fig. 4.6 Power at the maximum power point of two a-Si devices with different i-layer thicknesses (diamonds: 1000 nm; circles: 100 nm) at three illumination levels (1300, 1000 and 700 W m^{-2}). Figure created with data from Riesen et al. (2016)

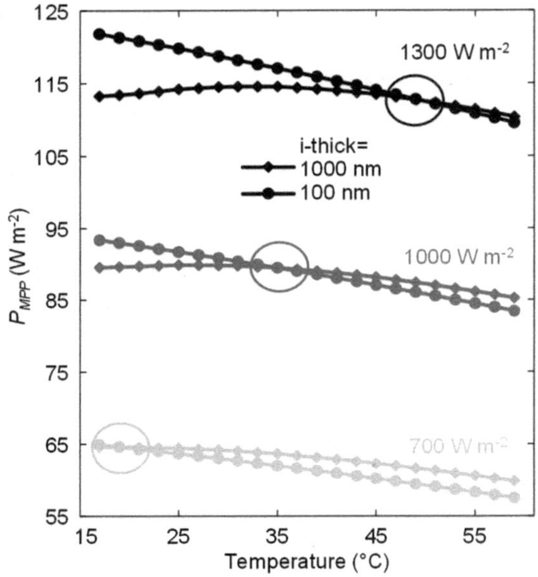

reported for perovskite solar cells have increased from 3.8% in 2009 (Kojima et al. 2009) up to over 22% in 2016 (NREL 2016).

The impact of temperature on the performances of perovskite based devices is investigated in several works (Aharon et al. 2014; Habisreutinger et al. 2014; Cojocaru et al. 2015; Ono et al. 2015; Zhang et al. 2015; Jacobsson et al. 2016a; Leong et al. 2016). The reported efficiencies, open-circuit voltages, fill factors and short-circuit current densities are shown in Fig. 4.7 as a function of temperature. The apparent discrepancies between the different results plotted in Fig. 4.7 can be attributed to several factors:

- the different architectures of the cells (planar or mesostructure),
- batch-to-batch and cell-to-cell variations which are known to be important for perovskite devices because their manufacturing is not entirely mastered yet,
- different device degradations that could be affected by the duration and protocol of the experiments. Perovskite devices are particularly sensitive and their performances can degrade rapidly over time. These degradations originate from several mechanisms including temperature-dependent processes (Misra et al. 2015). Perovskite cells also suffer from light-induced degradation but this effect seems to be reversible (Nie et al. 2016). Note that because of irreversible degradation issues, the data in Cojocaru et al. (2015), Jacobsson et al. (2016a) are measurements on two different devices for the temperature ranges above and below room temperature,
- differences in scan direction and speed for the electrical characterization. Indeed, one particularity of perovskite-based solar cells is their slow response to changes in voltage bias or illumination. This sometimes leads to a hysteresis between the

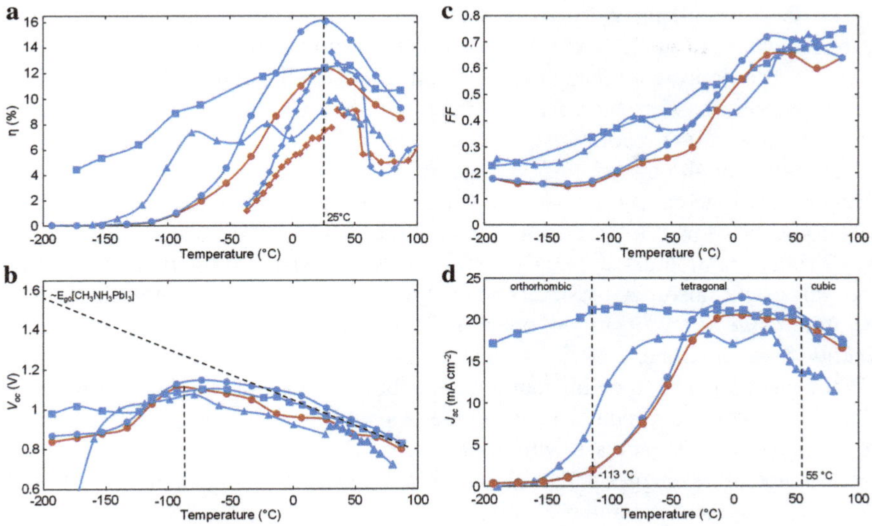

Fig. 4.7 Efficiency and IV parameters of different perovskite-based devices as a function of temperature. **a** Efficiency, **b** Open-circuit voltage, **c** Fill factor, **d** Short-circuit current density. Data sources: *squares* (Leong et al. 2016), *circles* (Zhang et al. 2015), *triangles* (Jacobsson et al. 2016a), *diamonds* (Cojocaru et al. 2015). Data plotted in *blue* were measured with a reverse scan and data plotted in *red* via a forward scan. Data measured on different devices are separated

IV curves measured in forward-bias and those measured in reverse-bias scans. This hysteresis and therefore the measured characteristics also depend on scan speed and device temperature (Ono et al. 2015; Jacobsson et al. 2016b). Several hypotheses have been proposed to explain the hysteretic behavior of perovskite devices: ferro-electric effects (Frost et al. 2014), ion migration (Tress et al. 2015), interfacial charge transfer (Jena et al. 2015) and capacitive effects (Kim and Park 2014). In any case, this behavior and the temperature and light induced degradations make it complicated to establish the actual steady-state performances of perovskite devices,

– differences in experimental set-up and protocol. For example, in Jacobsson et al. (2016) dry nitrogen gas is flushed on the temperature stage windows to minimize the formation of ice at low temperature and the measurements are corrected for the unavoidable ice layer.

Despite the discrepancies, the experimental results show that the efficiency of perovskite solar cells generally peaks around room temperature (Fig. 4.7a). It is not surprising that this maximum is found around room temperature since the devices are optimized for the standard test conditions (25 °C, AM1.5, 1000 W m^{-2}). Below this temperature, efficiency goes down more or less linearly depending on the studies. This thermal behavior is different from that of most other types of solar cells whose efficiencies increase with decreasing temperature (see Chap. 2). In all the studies, the efficiency drop with decreasing temperature is primarily driven by a strong decrease of the fill factor (Fig. 4.7c). This temperature dependence of the fill factor is also opposite to that common to other solar cell technologies where the fill factor decreases with increasing temperature because of a temperature-driven augmentation of the recombinations (see Sect. 2.3.3).

The measured open-circuit voltage of the perovskite cells shows a linear decrease with increasing temperature (Fig. 4.7b) similarly to that of other solar cell technologies (down to a certain temperature which is discussed below). The $V_{oc}(T)$ curves extrapolated linearly down to 0 K give values close to the bandgap of the perovskite absorber (dashed line in Fig. 4.7b). This means that, in the temperature range of PV operation, open-circuit voltage is not limited by non-optimum energy level alignment at the interfaces but results from the balance between photogeneration and recombination in the bulk of the perovskite (Leong et al. 2016). Since open-circuit voltage decreases with temperature, it signifies that recombination rates increase with temperature. Thus, the increase of the fill factor with temperature suggests that the limiting transport mechanisms are strongly dependent upon temperature. This could be explained by trap states in the perovskite layer (Leong et al. 2016). The important slopes of open-circuit voltage versus irradiance shown in Fig. 4a in Leong et al. (2016) support the hypothesis that trap-assisted recombinations dominate above 200 K. As temperature rises, trapped carriers are more likely to escape rapidly so the electric field becomes more uniform, favoring the transport and collection of photogenerated carriers. This would explain a positive temperature dependence of the fill factor. Additionally, as previously mentioned, the fill factor also depends on the recombination rates which increase with temperature. The result of this two opposite effects could explain the maximum sometimes observed for the fill factor (Fig. 4.7c).

Another interesting observation in these studies of temperature-dependent performances is that open-circuit voltage saturates and decreases below a temperature of about −80 °C (Fig. 4.7b). Different hypotheses on the physical origin of this trend are suggested in Jacobsson et al. (2016a), Leong et al. (2016). To determine the correct explanation, it will be necessary to separate the reversible influence of temperature from the irreversible degradations undergone by the devices during the experiments. Furthermore, it is critical to get rid of transient effects to be able to evaluate the actual steady-state performances of perovskite-based devices (Dunbar et al. 2016).

Another remarkable feature of perovskite compounds is that they can have a phase transition in the operating temperature range of photovoltaic operation. For example, $CH_3NH_3PbI_3$ undergoes a tetragonal to cubic transition around[5] 55 °C (Cojocaru et al. 2015). This transition has an impact on the bandgap and its temperature dependence but the existing measurements diverge (Dittrich et al. 2015; Jacobsson et al. 2016a; Leong et al. 2016). The temperature-dependent measurements from Cojocaru et al. (2015), the diamonds in Fig. 4.7, show a significant drop of V_{oc}, FF,

[5]Note that phase transition temperatures can be modified by constraints originating from the layers adjacent to the perovskite material (Dittrich et al. 2015; Jacobsson et al. 2016a).

J_{sc} and η around 55 °C. On the other hand, a minor boost in all these characteristics is observed in Jacobsson et al. (2016a), triangles in Fig. 4.7, around 60 °C. In the results from Leong et al. (2016), Zhang et al. (2015), the short-circuit current density decreases more rapidly above 55 °C. If the cubic and the tetragonal phases coexist on a temperature range, this drop in J_{sc} could be due to defects at the interface between the two phases (Leong et al. 2016). Below 55 °C, the short-circuit current tends to plateau and decreases at very low temperatures. In an ideal perovskite solar cell, the short-circuit current should increase with decreasing temperature. Indeed, a notable feature of perovskites is the reverse ordering of their band-edges states which results in a bandgap temperature dependence opposite to that of tetrahedrally coordinated semiconductors (see Sect. 2.2.2). This signifies that, at low temperatures, the short-circuit current is not limited by photogeneration but by the collection of photogenerated carriers.

It is fundamental for the PV community to acquire a detailed knowledge of the reversible temperature dependences of perovskite materials in order to:

- be able to separate these reversible effects from the permanent degradations undergone by the materials over time (which is complex because these degradations are also temperature dependent),
- predict the performances of perovskite devices in the field and calculate their potential energy yields and economic viability,
- improve the design of the devices for real operating conditions and in particular the design of tandem structures (especially monolithic devices which need to be current matched).

4.6 Multi-junction Solar Cells

The temperature sensitivity of a multi-junction device depends primarily on whether it is connected in series or in parallel. The temperature sensitivity of devices connected in parallel (such as four terminal tandem cells or certain spectrum splitting devices) is function of the separate temperature coefficients of all the sub-cells. On the other hand, for serially connected cells (the majority of multi-junction devices), the current is always limited by one of the sub-cells. Indeed, if the sub-cells are designed to be current matched, it is only for a given incident spectrum and a given temperature. This has some important consequences for the temperature coefficients of monolithic multi-junction solar cells.

Since the voltage of the whole device is equal to the sum of the voltages of the different sub-cells, its temperature sensitivity depends on that of all the sub-cells. However, since the current circulating through the device equals that of the limiting sub-cell, its temperature sensitivity depends on the temperature dependences of carrier generation and transport in the limiting sub-cell (Friedman 1996). This is illustrated in Fig. 4.8 which shows the fill factor temperature dependence of three tandem devices with different top-cell thicknesses. For one device, an abrupt

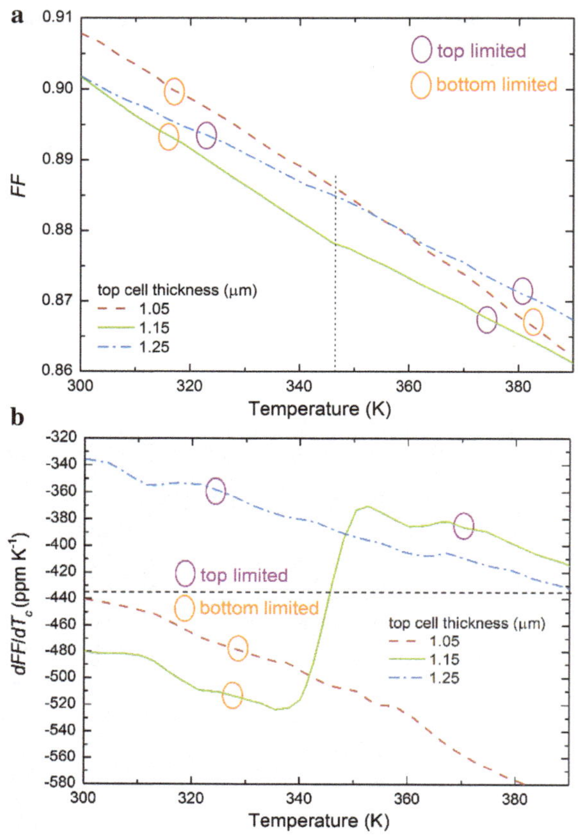

Fig. 4.8 **a** Fill factor and **b** fill factor temperature dependence, as a function of temperature for three tandem (GaInP/GaAS) devices with different top-cell thicknesses. Figure created with data from Friedman (1996)

change in the absolute temperature coefficient of the fill factor can be observed between 340 and 350 K (Fig. 4.8b). Note that a similar variation occurs for the temperature coefficient of short-circuit current in the same temperature range (Friedman 1996). These abrupt variations of the temperature coefficients correspond to the transition between a top-limited and a bottom-limited configuration. Indeed, because the semiconductors used for the different sub-cells have different bandgap temperature dependences (see Sect. 2.2.2), a good current matching can only be obtained at a certain temperature. This phenomenon has to be considered when modeling multi-junction devices. In particular, the temperature at which the device is current-matched should be tuned to correspond to the temperature at which the device will produce the most energy (see Sect. 1.2).

Also, multi-junction devices are designed to be current-matched for a certain spectrum (usually AM1.5). When the spectrum changes, the device becomes top or bottom limited (depending on if the change is a red or a blue shift). This modification of the limiting sub-cell has similar consequences to these due to a temperature change (i.e. the TCs of short-circuit current and fill factor change abruptly because they are driven by the dominant losses of one or the other sub-cell). This phenomenon is illustrated in Fig. 4 in Siefer and Bett (2012).

The observations that temperature and incident spectrum impact the performances and temperature coefficients of multi-junction devices illustrate why this kind of PV devices would especially benefit from design optimizations taking into account the actual operating conditions (instead of the standard test conditions).

Multi-junction solar cells, because they convert more efficiently the incident solar spectrum, have the benefit of significantly reduced heat sources (compared to single junction devices) which result in lower operating temperatures. However, in practice, because of their significant cost, multi-junction solar cells are mostly dedicated to concentrator applications where the incident radiation and thus the heat that needs to be evacuated is multiplied by the concentration factor.

4.7 Concentrator Photovoltaics

Photovoltaic systems working under concentration are known for their low temperature sensitivities. This originates from the increase in open-circuit voltage concomitant with concentration. Indeed, open-circuit voltage is an indicator of the generation-recombination balance in the cell, so it logically increases under concentration as there is more photogeneration. The open-circuit voltage can be written as a function of the concentration factor (*X*) as (see Sect. 2.2.1):

$$V_{oc} = \frac{kT_c}{q}\ln\left(\frac{X\,J_{sc,1sun}}{(1/ERE_{oc})\,J_{0,rad}}\right) = V_{oc,1sun} + \frac{kT_c}{q}(\ln(ERE_{oc}) + \ln(X)) \qquad (4.1)$$

where ERE_{oc} is the external radiative efficiency at open circuit and $J_{0,rad}$ is the dark current density in the radiative limit. Note that a linear behavior of J_{sc} with the concentration factor X is assumed. This assumption can fail if an important loss mechanism in short circuit has a significant intensity dependence. The temperature dependence of V_{oc} is (see Sect. 2.2.1):

$$\frac{dV_{oc}}{dT_c} = -\frac{\frac{E_{g0}}{q} - V_{oc} + \gamma\frac{kT_c}{q}}{T_c} \qquad (4.2)$$

with:

$$\gamma = 1 - \frac{d\ln ERE_{oc}}{d\ln T_c} + \left(2\frac{d\ln E_g}{d\ln T_c} - \frac{d\ln J_{sc,1sun}}{d\ln T_c}\right) \qquad (4.3)$$

These theoretical expressions show that concentrating sunlight upon a PV device improves its open-circuit voltage (4.1) and in turn improves its temperature coefficient (4.2). This is illustrated in Fig. 4.9 assuming an external radiative efficiency

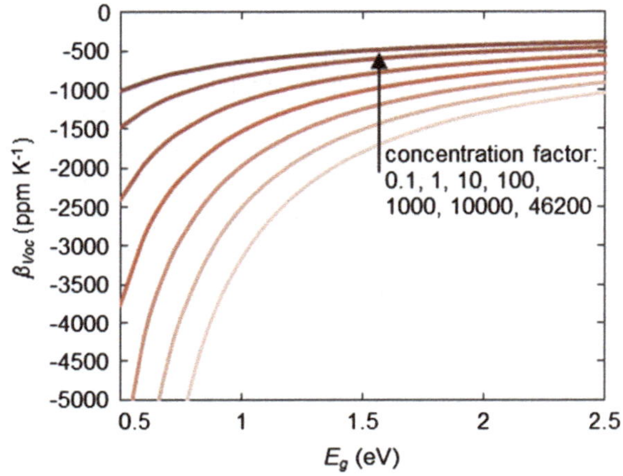

Fig. 4.9 Temperature coefficient of open-circuit voltage as a function of bandgap for different concentration factors (from 0.1 to 46,200) assuming an external radiative efficiency (ERE) equal to 10^{-3}

at open circuit of 10^{-3}. This behavior is experimentally confirmed for various systems under concentration (Yoon and Garboushian 1994; Braun et al. 2011; Landis et al. 2011; Siefer and Bett 2012). Also, all the recombination mechanisms are function of the excess carrier density which depends on photogeneration and thus on concentration. Consequently, the dominant recombination processes in a solar cell can change with the concentration factor. This impacts the temperature sensitivity of open-circuit voltage through the term $d \ln ERE_{oc}/d \ln T_c$ in (4.3).

The temperature sensitivity of the fill factor generally decreases with concentration following the behavior of the open-circuit voltage (see Sect. 2.3.3). The fill factor also increases with open-circuit voltage and thus with concentration. However, the effect of series resistance is magnified at high concentration because the current circulating through the cell is increased. Depending on cell design (contacts, metallization, etc.), this can lead to a drop of the fill factor above a certain concentration level (Siefer and Bett 2012).

As for the normalized temperature sensitivity of short-circuit current, it generally does not change significantly with concentration. This is due to the fact that the main drivers of the temperature coefficient of short-circuit current, i.e. the spectrum and the bandgap temperature sensitivities, do not depend upon the illumination level.

In addition to reducing the temperature sensitivity of PV devices, concentration has another notable effect on their thermal behavior: it increases the heat generated within the device by a factor close to that of concentration (in fact, it is a little less because the conversion efficiency increases with concentration which reduces the fraction of incident power that contributes to heating the device). This means that design approaches that take into account the heat source are particularly important for concentrated photovoltaics. For example, using a spectrally selective reflector enables to concentrate only the photons above the bandgap onto the device and eliminate the heat source generated by the parasitic absorption of sub-bandgap photons. It is shown in Maghanga et al. (2011) that such a design can lead to reductions in operating temperature of more than 20 °C at a concentration factor of 25.

PV systems with a concentration factor above a few dozen are generally equipped with active cooling devices to maintain the cell temperature at an acceptable level. Ideally, one would want to have an operating temperature as low as possible but the additional impact of the cooling apparatus, on both energetic efficiency and cost, has to be taken into consideration.

4.8 Thermophovoltaic Devices

Thermophotovoltaics (TPV) refers to the conversion of radiation from thermal sources into electricity [Coutts and Fitzgerald 1998; Bauer 2011; Bermel, Sect. 15 in Boriskina et al. (2016)]. There are several differences with conventional solar photovoltaics. First, the heat source is at a lower temperature than the equivalent blackbody temperature of the Sun ($\sim$5800 K). In TPV, the illumination comes from a body called radiator (or emitter) at temperatures usually smaller than 2500 K (Fig. 4.10). Instead of high energy photons from visible light, lower energy photons from the infrared spectral range are involved. There exist several kinds of TPV systems: in standard systems, waste heat sources, combustion or nuclear reactions supply the thermal energy to the radiator; in solar thermophotovoltaics (STPV), concentrated solar radiation is used to heat up the radiator via an absorber layer on top of the emitter (Swanson 1979; Demichelis and Minetti-Mezzetti 1980); in thermophotonics (TPX), an electrochemical potential is applied to the radiator—typically a light emitting diode attached to the emitter—in order to enhance the illumination toward the cell (Harder and Green 2003); in near-field radiation thermophotovoltaics (NFR-TPV), the radiator and the PV cell are separated by a distance inferior to the characteristic wavelength of thermal radiation, given by Wien's law, such that thermal radiation transfer is increased by orders of

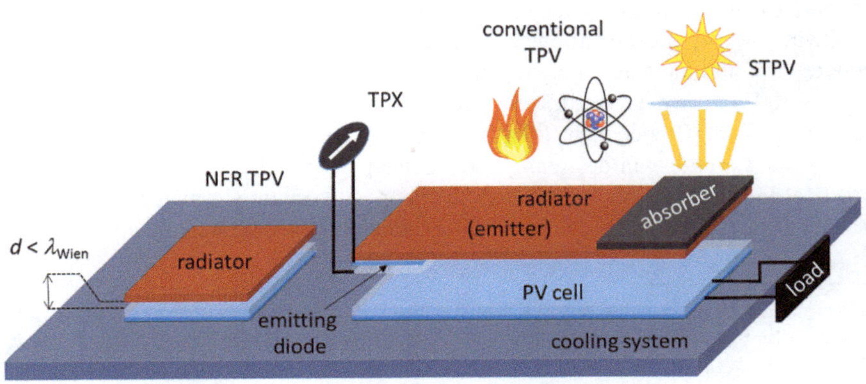

Fig. 4.10 Schematic of standard thermophotovoltaic (TPV), solar thermophovoltaic (STPV), thermophotonic (TPX), and near-field radiation thermophotovoltaic (NFR TPV) systems

magnitude above blackbody radiation due to the additional contribution of evanescent waves (Whale and Cravalho 2002; Narayanaswamy and Chen 2003; Greffet, Sect. 16 in Boriskina et al. 2016).

Because the average energy of the photons in TPV systems is lower than that of solar illumination, the semiconductors efficient for TPV have smaller bandgaps than those used for solar PV. Typically, TPV cells are made of Ge (0.66 eV) or from group IV or III–V compounds (e.g. GaSb (0.72 eV), $In_xGa_{1-x}As_ySb_{1-y}$ (0.5 to 0.6 eV), InAs (0.36 eV),…).

Unlike non-concentrated solar illumination, radiation is usually incident on the TPV cell over the full hemisphere. The first consequence is that a planar blackbody radiator at 91 °C is sufficient to generate an illumination equal to that of the AM1.5 solar irradiation (1000 W m^{-2}).[6] The emission of a planar blackbody radiator at 2000 K corresponds to a solar illumination concentrated 907 times. It follows that single junction TPV cells generate output power densities of the order of the watt per square centimeter (Teofilo et al. 2008; Bauer 2011), which is significantly larger than that of the best non-concentrator solar cells.[7] Another consequence of the hemispherical illumination is that the so-called angle mismatch loss (Sect. 2.1) can be avoided by making the channels for radiation from the emitter to the cell and from the cell to the emitter fully reciprocal. Suppression of this fundamental loss has implications on the bandgap-voltage offset—thus on open-circuit voltage—, and on the temperature coefficient. One additional advantage, not widely accounted for, is that the external electroluminescence from the cell towards the radiator, which is in the spectral range where the radiator emits/absorbs, can be recycled.

In TPV systems, the heat generation due to the absorption of sub-bandgap photons (which are useless for photovoltaic conversion) can be avoided (Harder and Würfel 2003). This can be achieved by designing spectrally selective radiators/emitters (Coutts and Fitzgerald 1998; Zhou et al. 2016; Pfiester et Vandervelde 2016), or by using reflectors that can be either a filter placed between the radiator and the cell or layers added at the front or the back of the cell (Bauer 2011). One advantage in TPV is that this energy useless and detrimental to the PV converter can be externally recycled by the radiator.

Owing to the aforementioned advantages, an ideal solar thermophotovoltaic converter would have an efficiency of 85% at maximum concentration (Harder and Würfel 2003). However, the best efficiencies of STPV systems to date are of the order of a few percents (Lenert et al. 2014; Ungaro et al. 2015; Bierman et al. 2016). Despite the constant progress in designing and fabricating optically efficient structures, the practical implementation of the ideal absorbers and emitters remains a challenging issue: out of 20.3 W (resp. 25 W) of solar radiation incident upon the absorber, only

---

[6]However, a PV cell operating at 300 K would not be very efficient just because of the Carnot loss limitation (in that case η_{Carnot} = 1–300/364.42 = 17.7 %). The cell would need to be cooled (e.g. at T_c = 77 K, η_{Carnot} = 78.9 %).

[7]According to the NREL chart of the best-research cell efficiencies (NREL 2016), non-concentrator cell efficiencies are currently below 40%, which corresponds to less than 0.04 W cm^{-2} power output for an AM1.5 illumination.

29% (resp. 16%) reaches the PV cell in Lenert et al. (2014) [resp. in. Ungaro et al. (2015)]. Performances of the TPV part (emitter—PV cell) are currently limited as well: the largest measured cell efficiencies of standard TPV converters are 20–25% at 25° C (Wernsman et al. 2004; Crowley et al. 2005; Wernsman et al. 2005; Teofilo et al. 2008).

The thermal behavior of TPV devices has an important impact on their performances. The high-temperature radiator is close to the PV cells thus creating severe thermal conditions: a large temperature gradient must be maintained without much possibilities for extracting heat from the device at its front surface. A cooling system is required at the back surface to prevent an excessive temperature rise (Bauer 2011). Actual TPV cells are sensitive to temperature in the same way solar cells are (see Chaps. 1 and 2): conversion efficiency decreases with temperature, at rates ranging from −0.1 to nearly −1% K^{-1} according to Moriarty and Emery (1998). Different studies report on the temperature coefficient of the power output normalized at 25 °C for various TPV cells: around −0.5% K^{-1} for a GaSb cell in Burger and Mueller (1995), −0.46 to −0.5% K^{-1} for a GaInAs cell in Moriarty and Emery (1998), −0.41% for a GaSb cell in Martin et al. (2004). This decrease is mainly due to negative temperature sensitivities of open-circuit voltage and then of the fill factor. The temperature coefficient of short-circuit current is usually positive: between 0.1 and 0.25% K^{-1} for a GaSb cell in Burger and Mueller (1995), 0.26% K^{-1} for a GaSb cell in Martin et al. (2004), but slightly negative values are reported in Moriarty and Emery (1998) for a GaInAs cell. In Teofilo et al. (2008), an inventory for a set of TPV cells dedicated to space applications indicates a trend: the magnitude of the temperature coefficient of efficiency decreases when efficiency improves (see Fig. 4.11). Variations of temperature coefficients of open-circuit voltage with illumination (varying with the temperature of blackbody cavities or using variable high-intensity lamps) are similar to those reported for concentrator photovoltaics (see Sect. 4.7) and consistent with theoretical expectations: the magnitude of the negative temperature coefficient of open-circuit voltage decreases

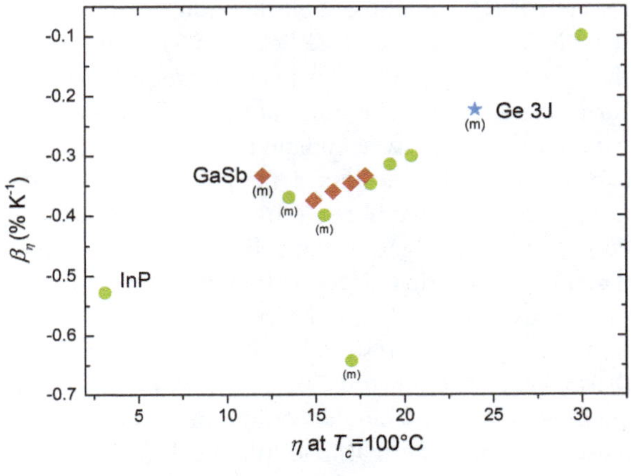

Fig. 4.11 Temperature coefficient of efficiency, as a function of efficiency interpolated at 100 °C, for InP, GaSb single-junction and Ge triple-junction TPV cells. Labels (m) indicate the data from measurements. Figure created from data in Teofilo et al. (2008)

when the illumination level increases. This is shown experimentally for an InGaAs TPV cell in Moriarty and Emery (1998) and for InGaAsSb, GaSb and Ge TPV cells in Sulima et al. (2002). However, there is no consensus on the trends for the magnitude of the negative temperature coefficient of the fill factor (Moriarty and Emery 1998; Sulima et al. 2002). The results suggest that there might be peculiar behaviors of temperature sensitivities with the illumination level for TPV cells. Thus, one should be cautious when comparing temperature coefficients from different studies. It is worth mentioning that while illumination is normalized for solar cells (with the reference AM1.5 spectrum), for thermophotovoltaics, unless blackbodies are used, much variability is expected for the spectral distribution and the total level of radiation from different emitters. As mentioned previously, one advantage of thermophotovoltaics is the opportunity to design and build spectrally selective emitters to maximize photovoltaic conversion efficiency. This asset removes the interest of a normalization of illumination for performing comparative analyses between different studies: the use of blackbody radiators would imply inequalities in the mitigation of losses among different TPV cell technologies.

Accounting for the thermal behaviour of PV cells has an impact on the optimum design parameters of TPV systems. In DeSutter et al. (2016), a genetic algorithm is used to determine the spectrally selective emitters that maximize the performances of a GaSb TPV cell for a specific cooling system. The emissivity of a radiator at 2000 K that maximizes power output corresponds to the filtering of both the sub-bandgap photons (see Sect. 3.5.1) and high-energy photons (see Sect. 3.5.2). Usually a step emissivity function with a cut-off at the band-gap of the cell is targeted when spectrally selective emitters are designed. The simulations made in DeSutter et al. (2016) show that a rectangular function yields more electrical power output when taking into consideration heat generation within the cell, the equilibrium temperature reached with a given cooling system, and thus the efficiency loss associated with the cell temperature (see Chap. 3). In addition to avoiding the sub-bandgap heat source, removing high-energy photons can be beneficial, even though the photocurrent is decreased, because the thermalization heat source is reduced. As for the optimal low-energy cut-off, it corresponds to the bandgap at the equilibrium temperature of the cell. The same kind of optimum filtering is demonstrated in Bernardi et al. (2015) for a NFR-TPV device (see Sect. 3.5.2). In the case of STPV systems, the coupled thermal equilibria of both the absorber/emitter and the PV cell have to be accounted for in the analysis. The optimum cut-off for the absorptivity of the absorber results from a trade-off between a gain by absorption of the incident solar irradiation and a loss by thermal emission at the equilibrium temperature of the absorber/emitter (Dupré et al. 2016). If a step emissivity function is assumed for the emitter with a cut-off at the bandgap of the cell (emissivity equal to one above, zero below), then the optimum bandgap is a function of the performances of the cooling system and of the quality of the cell (characterized by its External Radiative Efficiency, see Sect. 3.5.3). The previous results are illustrations of TPV device designs driven by the thermal criteria defined in Sect. 3.5 (minimizing the heat source and the temperature coefficient, maximizing the cooling performances), with an interplay of the different losses that are a thermal load to the cell (Sect. 3.2).

References

Aharon S, Dymshits A, Rotem A, Etgar L (2014) Temperature dependence of hole conductor free formamidinium lead iodide perovskite based solar cells. J Mater Chem A 3:9171–9178. Doi:10.1039/C4TA05149A

Bauer T (2011) Thermophotovoltaics: basic principles and critical aspects of system design. Springer

Bernardi MP, Dupré O, Blandre E et al (2015) Impacts of propagating, frustrated and surface modes on radiative, electrical and thermal losses in nanoscale-gap thermophotovoltaic power generators. Sci Rep. Doi:10.1038/srep11626

Bierman DM, Lenert A, Chan WR et al (2016) Enhanced photovoltaic energy conversion using thermally based spectral shaping. Nat Energy 1:16068. Doi:10.1038/nenergy.2016.68

Boriskina SV, Green MA, Catchpole K et al (2016) Roadmap on optical energy conversion. J Opt 18:73004. Doi:10.1088/2040-8978/18/7/073004

Braun A, Hirsch B, Vossier A (2011) Temperature dynamics of multijunction concentrator solar cells up to ultra high irradiance. Prog Photovoltaics Res Appl 2–8. Doi:10.1002/pip

Burger DR, Mueller RL (1995) Characterization of thermophotovoltaic cells. AIP Conf Proc 321:457–472. Doi:10.1063/1.47054

Carlson DE, Lin G, Ganguly G (2000) Temperature dependence of amorphous silicon solar cell PV parameters. In: Twenty-Eighth IEEE photovoltaic specialists conference

Cojocaru L, Uchida S, Sanehira Y et al (2015) Temperature effects on the photovoltaic performance of planar structure perovskite solar cells. Chem Lett 1557–1559. Doi:10.1246/cl.150781

Coutts T, Fitzgerald M (1998) Thermophotovoltaics. Sci Am 90–95

Crowley CJ, Elkouh NA, Murray S, Chubb DL (2005) Thermophotovoltaic converter performance for radioisotope power systems. AIP Conf Proc 746:601–614. Doi:10.1063/1.1867178

Demichelis F, Minetti-Mezzetti E (1980) A solar thermophotovoltaic converter. Sol Cells 1:395–403. Doi:10.1016/0379-6787(80)90063-0

DeSutter J, Bernardi MP, Francoeur M (2016) Determination of thermal emission spectra maximizing thermophotovoltaic performance using a genetic algorithm. Energy Convers Manag 108:429–438. Doi:10.1016/j.enconman.2015.11.029

De Wolf S, Descoeudres A, Holman ZC, Ballif C (2012) High-efficiency silicon heterojunction solar cells: a review. Green 2:7–24. Doi:10.1515/green-2011-0039

Dittrich T, Awino C, Prajongtat P et al (2015) Temperature dependence of the band gap of CH3NH3PbI3 stabilized with PMMA: a modulated surface photovoltage study. J Phys Chem C 119:23968–23972. Doi:10.1021/acs.jpcc.5b07132

Dunbar RB, Moustafa W, Pascoe AR (2016) Device pre-conditioning and steady-state temperature dependence of $CH_3NH_3PbI_3$ perovskite solar cells. Prog. Photovolt. Res. Appl. Doi: 10.1002/pip.2839

Dupré O (2015) Physics of the thermal behavior of photovoltaic devices. Ph.D. Thesis. INSA Lyon, France

Dupré O, Vaillon R, Green MA (2015) Experimental assessment of temperature coefficient theories for silicon solar cells. IEEE J Photovoltaics 1–5. Doi:10.1109/JPHOTOV.2015.2489864

Dupré O, Vaillon R, Green MA (2016) Optimization of solar thermophotovoltaic systems including the thermal balance. In: Proceedings of the 43rd IEEE PVSC

Fischer D, Bichsel F, Bruijn S De, et al (2009) Positive effective temperature coefficient of power of +0.75%/°C in flexible a-Si modules in building integrated installations. In: 24th European photovoltaic solar energy conference, pp 3505–3508

Friedman DJ (1996) Modelling of tandem cell temperature coefficients. In: 25th IEEE photovoltaic specialists conference. IEEE, pp 89–92

Frost JM, Butler KT, Walsh A (2014) Molecular ferroelectric contributions to anomalous hysteresis in hybrid perovskite solar cells. APL Mater. Doi:10.1063/1.4890246

Green MA, Emery K, Blakers AWW (1982) Silicon solar cells with reduced temperature sensitivity. Electron Lett 18:97–98. Doi:10.1049/el:19820066

Green MA, Blakers AW, Osterwald CR (1985) Characterization of high-efficiency silicon solar cells. J Appl Phys 58:4402. Doi:10.1063/1.336286

Green MA, Ho-Baillie A, Snaith HJ (2014) The emergence of perovskite solar cells. Nat Photonics 8:506–514. Doi:10.1038/nphoton.2014.134

Habisreutinger SN, Leijtens T, Eperon GE et al (2014) Carbon nanotube/polymer composites as a highly stable hole collection layer in perovskite solar cells. Nano Lett 14:5561–5568. Doi:10.1021/nl501982b

Harder N, Green MA (2003) Thermophotonics. Semicond Sci Technol 18:S270–S278

Harder N, Würfel P (2003) Theoretical limits of thermophotovoltaic solar energy conversion. Semicond Sci Technol 151:S151–S157. Doi:10.1088/0268-1242/18/5/303

Jacobsson TJ, Tress W, Correa-Baena J-P, et al (2016a) Room Temperature as a goldilocks environment for CH_3 NH_3 PbI_3 perovskite solar cells: the importance of temperature on device performance. J Phys Chem C acs.jpcc.6b02858. Doi:10.1021/acs.jpcc.6b02858

Jacobsson TJ, Tress W, Correa-Baena J-P, et al (2016b) Supporting information: Room temperature as a goldilocks environment for $CH_3NH_3PbI_3$ perovskite solar cells: the importance of temperature on device performance. J Phys Chem C 46:acs.jpcc.6b02858. Doi: 10.1021/acs.jpcc.6b02858

Jena AK, Chen HW, Kogo A et al (2015) The interface between FTO and the TiO_2 compact layer can be one of the origins to hysteresis in planar heterojunction perovskite solar cells. ACS Appl Mater Interfaces 7:9817–9823. Doi:10.1021/acsami.5b01789

Kim HS, Park N-G (2014) Parameters affecting I–V hysteresis of CH_3 NH_3 PbI_3 perovskite solar cells: effects of perovskite crystal size and mesoporous TiO_2 layer. J Phys Chem Lett 140811175830000. Doi:10.1021/jz501392m

Kojima A, Teshima K, Shirai Y, Miyasaka T (2009) Organometal halide perovskites as visible-light sensitizers for photovoltaic cells. J Am Chem Soc 131:6050–6051. Doi:10.1021/ja809598r

Kondo M, Nishio H, Kurata S et al (1997) Effective conversion efficiency enhancement of amorphous silicon modules by operation temperature elevation. Sol Energy Mater Sol Cells 49:1–6. Doi:10.1016/S0927-0248(97)00168-2

Landis GA., Belgiovane DJ, Scheiman DA (2011) Temperature coefficient of multijunction space solar cells as a function of concentration. 2011 37th IEEE photovolt specialists conference 001583–001588. Doi:10.1109/PVSC.2011.6186260

Lenert A, Bierman DM, Nam Y et al (2014) A nanophotonic solar thermophotovoltaic device (supplementary information). Nat Nanotechnol 9:126–130. Doi:10.1038/nnano.2013.286

Leong WL, Ooi Z-E, Sabba D et al (2016) Identifying fundamental limitations in halide perovskite solar cells. Adv Mater 28:2439–2445. Doi:10.1002/adma.201505480

Maghanga CM, Niklasson GA, Granqvist CG, Mwamburi M (2011) Spectrally selective reflector surfaces for heat reduction in concentrator solar cells: modeling and applications of TiO_2: Nb-based thin films. Appl Opt 50:3296–302. Doi:10.1364/AO.50.003296

Makrides G, Zinsser B, Norton M, Georghiou G (2012a) Performance of Photovoltaics under actual operating conditions. In: Fthenakis DV (ed) Third generation photovoltaics. InTech, pp 201–232

Makrides G, Zinsser B, Phinikarides A et al (2012b) Temperature and thermal annealing effects on different photovoltaic technologies. Renew Energy 43:407–417. Doi:10.1016/j.renene.2011.11.046

Martin D, Algora C, Martín D, Algora C (2004) Temperature-dependent GaSb material parameters for reliable thermophotovoltaic cell modelling. Semicond Sci Technol 19:1040–1052. Doi:10.1088/0268-1242/19/8/015

Mishima T, Taguchi M, Sakata H, Maruyama E (2011) Development status of high-efficiency HIT solar cells. Sol Energy Mater Sol Cells 95:18–21. Doi:10.1016/j.solmat.2010.04.030

Misra RK, Aharon S, Li B et al (2015) Temperature- and component-dependent degradation of perovskite photovoltaic materials under concentrated sunlight. J Phys Chem Lett 6:326–330. Doi:10.1021/jz502642b

Moriarty T, Emery K (1998) Thermophotovoltaic cell temperature measurement issues. In: Proceedings of the 4th NREL conference on thermophotovoltaic generation of electricity

Moser D, Pichler M, Nikolaeva-Dimitrova M (2013) Filtering procedures for reliable outdoor temperature coefficients in different photovoltaic technologies. J Sol Energy Eng 136:1–10. Doi:10.1115/1.4024931

Narayanaswamy A, Chen G (2003) Surface modes for near field thermophotovoltaics. Appl Phys Lett 82:3544–3546. Doi:10.1063/1.1575936

Nie W, Blancon J, Neukirch AJ, et al (2016) Light-activated photocurrent degradation and self-healing in perovskite solar cells. Nat Commun 7:1–9. Doi:10.1038/ncomms11574

NREL (2016) PV record efficiency chart from NREL. http://www.nrel.gov/ncpv/. Accessed 18 May 2016

Ono LK, Raga SR, Wang S et al (2015) Temperature-dependent hysteresis effects in perovskite-based solar cells. J Mater Chem A 3:9074–9080. Doi:10.1039/C4TA04969A

Pfiester NA, Vandervelde TE (2016) Selective emitters for thermophotovoltaic applications. Physica Status Solidi a 1–24. https://doi.org/10.1002/pssa.201600410

Ponce-Alcántara S, Connolly JP, Sánchez G et al (2014) A statistical analysis of the temperature coefficients of industrial silicon solar cells. Energy Procedia 55:578–588. Doi:10.1016/j.egypro.2014.08.029

Richter A, Hermle M, Glunz SW (2013) Reassessment of the limiting efficiency for crystalline silicon solar cells. IEEE J Photovoltaics 3:1184–1191. Doi:10.1109/JPHOTOV.2013.2270351

Riesen YS (2016) Energy yield and electricity management of thin-film and crystalline silicon solar cells. Ph.D. Thesis. EPFL, Switzerland

Riesen Y, Stuckelberger M, Haug F-J et al (2016) Temperature dependence of hydrogenated amorphous silicon solar cell performances. J Appl Phys 119:44505. Doi:10.1063/1.4940392

Seif JP, Descoeudres A, Filipic M et al (2014) Amorphous silicon oxide window layers for high-efficiency silicon heterojunction solar cells. J Appl Phys. Doi: 10.1063/1.4861404

Seif JP, Krishnamani G, Ballif C, De Wolf S (2015) Amorphous/crystalline silicon interface passivation: ambient-temperature dependence and implications for solar cell performance. IEEE J Photovoltaics 5:718–724

Siefer G, Bett AW (2012) Analysis of temperature coefficients for III–V multi-junction concentrator cells. Prog Photovoltaics Res Appl 22:515–524. Doi:10.1002/pip

Silverman TJ, Jahn U, Friesen G et al (2014) Characterisation of performance of thin-film photovoltaic technologies (from Report IEA-PVPS T13-02:2014)

Sulima OV, Bett AW, Dutta PS et al (2002) GaSb-, InGaAsSb-, InGaSb-, InAsSbP- and Ge-TPV cells with diffused emitters. In: Proceedings of the 29th IEEE Photovoltaic specialists conference 2002:892–895. Doi:10.1109/PVSC.2002.1190723

Swanson RM (1979) A proposed thermophotovoltaic solar energy conversion system. Proc IEEE 67:446–447. Doi:10.1109/PROC.1979.11270

Sze S, Ng K (1981) Physics of semiconductor devices. Wiley, New York

Taguchi M, Maruyama E, Tanaka M (2008) Temperature dependence of amorphous/crystalline silicon heterojunction solar cells. Jpn J Appl Phys 47:814–818. Doi:10.1143/JJAP.47.814

Tanay F, Dubois S, Enjalbert N, Veirman J (2011) Low temperature-coefficient for solar cells processed from solar-grade silicon purified by metallurgical route. Prog Photovoltaics Res Appl 19:966–972. Doi:10.1002/pip.1104

Tayyib M, Odden JO, Saetre TO (2013) UV-induced degradation study of multicrystalline silicon solar cells made from different silicon materials. Energy Procedia 38:626–635. Doi:10.1016/j.egypro.2013.07.326

Tayyib M, Odden JO, Saetre TO (2014a) Irradiance dependent temperature coefficients for MC solar cells from elkem solar grade silicon in comparison with reference polysilicon. Energy Procedia 55:602–607. Doi:10.1016/j.egypro.2014.08.032

Tayyib M, Odden JO, Saetre TO (2014b) Influence of spectral composition on the temperature coefficients of solar cells from elkem solar. In: Proceedings of the 40th IEEE Photovoltaic specialist conference: 2630–2634

Teofilo VL, Choong P, Chang J et al (2008) Thermophotovoltaic energy conversion for space. J Phys Chem C 112:7841–7845. Doi:10.1021/jp711315c

Tress W, Marinova N, Moehl T et al (2015) Understanding the rate-dependent J-V hysteresis, slow time component, and aging in CH_3 NH_3 PbI_3 perovskite solar cells: the role of a compensated electric field. Energy Environ Sci 8:995–1004. Doi:10.1039/C4EE03664F

Ungaro C, Gray SK, Gupta MC (2015) Solar thermophotovoltaic system using nanostructures. Opt Express 23:A1149. Doi:10.1364/OE.23.0A1149

Van Cleef MWM, Rubinelli FA, Rizzoli R, et al (1998) Amorphous silicon carbide/crystalline silicon heterojunction solar cells: a comprehensive study of the photocarrier collection. Jpn J Appl Phys Part 1 Regul Pap Short Notes Rev Pap 37:3926–3932. Doi:10.1143/JJAP.37.3926

Virtuani A, Pavanello D, Friesen G (2010) Overview of temperature coefficients of different thin film photovoltaic technologies. In: 25th European photovoltaic solar energy conference

Wernsman B, Siergiej RR, Link SD et al (2004) Greater than 20% radiant heat conversion efficiency of a thermophotovoltaic radiator/module system using reflective spectral control. IEEE Trans Electron Devices 51:512–515. Doi:10.1109/TED.2003.823247

Wernsman B, Mahorter RG, Siergiej R et al (2005) Advanced thermophotovoltaic devices for space nuclear power systems. AIP Conf Proc 746:1441–1448. Doi:10.1063/1.1867275

Whale MD, Cravalho EG (2002) Modeling and performance of microscale thermophotovoltaic energy conversion devices. IEEE Trans Energy Convers 17:130–142. Doi:10.1109/60.986450

Xiao C, Yu X, Yang D, Que D (2014) Impact of solar irradiance intensity and temperature on the performance of compensated crystalline silicon solar cells. Sol Energy Mater Sol Cells 128:427–434. Doi:10.1016/j.solmat.2014.06.018

Yoon S, Garboushian V (1994) Reduced temperature dependence of high-concentration photovoltaic solar cell open-circuit voltage (Voc) at high concentration levels. In: 1st World Conference on Photovoltaic Energy Conversion—WCPEC (a joint conference of PVSC, PVSEC and PSEC). IEEE, pp 1500–1504

Zhang H, Qiao X, Shen Y et al (2015) Photovoltaic behaviour of lead methylammonium triiodide perovskite solar cells down to 80 K. J Mater Chem A 3:11762–11767. Doi:10.1039/C5TA02206A

Zhao J, Wang A, Robinson SJ, Green MA (1994) Reduced temperature coefficients for recent high performance silicon solar cells. Prog Photovoltaics Res Appl 2:221–225. Doi:10.1002/pip.4670020305

Zhou Z, Sakr E, Sun Y, Bermel P (2016) Solar thermophotovoltaics: reshaping the solar spectrum. Nanophotonics. Doi:10.1515/nanoph-2016-0011

Index

O. Dupré et al., *Thermal Behavior of Photovoltaic Devices*,
DOI 10.1007/978-3-319-49457-9

Kolibri 360 GmbH
Ferdinand-Jühlke-Straße 7
99095 Erfurt, Deutschland
produktsicherheit@kolibri360.de